Fullerine Nano Materials

(Symposium of IUMRS-ICA 2008)

Journal of Physics: Conference Series Volume 159

Nagoya, Japan
9-13 December 2008

ISBN: 978-1-61738-375-5

Printed from e-media with permission by:

Curran Associates, Inc.
57 Morehouse Lane
Red Hook, NY 12571
www.proceedings.com

Some format issues inherent in the e-media version may also appear in this print version.

Printed by Curran Associates, Inc. (2010)

Institute of Physics
Dirac House, Temple Back
Bristol BS1 6BE UK

Phone: 44 1 17 929 7481
Fax: 44 1 17 920 0979

techtracking@iop.org

TABLE OF CONTENTS

INVITED TALKS

CONTRIBUTED TALKS

POSTERS

PREFACE

This volume contains peer-reviewed invited and contributed papers that were presented in Symposium N 'Fullerene Nano Materials' at the IUMRS International Conference in Asia 2008 (IUMRS-ICA 2008), which was held on 9–13 December 2008, at Nagoya Congress Center, Nagoya, Japan.

Over twenty years have passed since the discovery of C60 in 1985. The discovery of superconductivity of C60 in 1991 suggested infinite possibilities for fullerenes. On the other hand, a new field of nanocarbon has been developed recently, based on novel functions of the low-dimensional fullerene nanomaterials that include fullerene nanowhiskers, fullerene nanotubes, fullerene nanosheets, chemically modified fullerenes, endohedral fullerenes, thin films of fullerenes and so forth. Electrical, electrochemical, optical, thermal, mechanical and various other properties of fullerene nanomaterials have been investigated and their novel and anomalous nature has been reported. Biological properties of fullerene nanomaterials also have been investigated both in medical applications and toxicity aspects.

The recent research developments of fullerene nanomaterials cover a variety of categories owing to their functional diversity. This symposium aimed to review the progress in the state-of-the-art technology based on fullerenes and to offer the forum for active interdisciplinary discussions. 24 oral papers containing 8 invited papers and 22 poster papers were presented at the two-day symposium. Topics on the social acceptance of nanomaterials including fullerene were presented on the first day of the symposium. Biological impacts of nanomaterials and the importance of standardization of nanomaterials characterization were also shown. On the second day, the synthesis, properties, functions and applications of various fullerene nanomaterials were shown in both the oral and poster presentations.

We are grateful to all invited speakers and many participants for valuable contributions and active discussions. This symposium was partly supported by the Coordination Program of Science and Technology Projects "Developing Nanotechnologies and Engaging the Public" conducted by the Council for Science and Technology Policy (CSTP) and funded by Special Coordination Funds for Promoting Science and Technology.

ORGANIZING COMMITTEE OF SYMPOSIUM

(IUMRS-ICA 2008)

Chair

Kun'ichi Miyazawa (National Institute for Materials Science, Japan)

Vice Chairs

Daisuke Fujita (National Institute for Materials Science, Japan)
Takatsugu Wakahara (National Institute for Materials Science, Japan)
Tokushi Kizuka (Tsukuba University, Japan)
Kiyoto Matsuishi (Tsukuba University, Japan)
Yuichi Ochiai (Chiba University, Japan)
Masaru Tachibana (Yokohama City University, Japan)
Hironori Ogata (Hosei University, Japan)
Tadahiko Mashino (Keio University, Japan)
Ryotaro Kumashiro (Tohoku University, Japan)
Hidetoshi Oikawa (Tohoku University, Japan)

Journal of Physics: Conference Series **159** (2009) 012001 doi:10.1088/1742-6596/159/1/012001

Current standardization activities for the measurement and characterization of nanomaterials and structures

Shingo Ichimura[1], **Hiroshi Itoh**[1] **and Toshiyuki Fujimoto**[1]

[1]National Institute of Advanced Industrial Science and Technology (AIST), Umezono 1-1-1, Tsukuba, Ibaraki 305-8568, Japan

E-mail: s.ichimura@aist.go.jp

Abstract. Current standardization activities relating to the measurement and characterization of nanomaterials and nanostructures are described. The working items relating to single-wall and multiwall carbon nanotubes, which have been discussed by WG2 of ISO TC229 on Nanotechnology, are explained together with the social requests for standardization and the scope and structure of ISO TC229. In addition, the standardization of AFM probe characterization, which is under discussion at ISO TC201 (Surface Chemical Analysis)/SC9 (Scanning Probe Microscopy), is introduced. A new type of probe characterizer necessary for the precise measurement of probe shape is described, and the merit and utility of a probe-shape function that gives the relation between the probe length and width are also explained.

1. Introduction

After the establishment of the National Nanotechnology Initiative (NNI) by the USA in 2000, many countries have accelerated their R&D programs for nanotechnologies. The amount of governmental investment for nanotechnology R&D in 2005, for example, ran over 40 billion US$, which corresponds to a fivefold increase over the investment in 2000 [1]. In accordance, various nanomaterials and structures have been fabricated and utilized with the aim of creating new functions. Nanotechnology is expected to be a powerful driving force for industry in the 21st century. To encourage and promote its development, a trial has begun in which a special mark is placed on products fabricated by nanotechnology [2]. However, a negative social response toward nanotechnology has appeared based on a fear that engineered nanomaterials might be harmful to human health [3]. Thus, the standardization of nanotechnology has become an important and urgent issue in offering a scientifically sound basis for both supporting the positive aspects and examining the negative impact of nanotechnology.

The standardization of nanotechnology was first initiated as actions on national or regional levels in the USA, EU, and Asia. These were followed by the International Organization for Standardization (ISO) Technical Committee (TC) 229 on "Nanotechnology," which started activities in November 2005. This committee currently has 4 working groups (WG), among which WG2 is for "Measurement and Characterization" of nanotechnology. WG2 is also supported as a joint activity (JWG2) by the International Electrotechnical Commission (IEC) TC113 on "Nanotechnology Standardization for Electrical and Electronic Products and Systems."

The activities of ISO TC229 have stimulated discussion on nanotechnology standardization among various organizations concerned with standardization. In Japan, the practical need for nanotechnology standardization was investigated through a questionnaire survey of stakeholders in nanotechnology businesses. Figure 1 shows the results of the survey on key factors and sizes for the realization of new functions by using nanotechnology [4]. It is clear from the results that most of the stakeholders surveyed recognize the importance of shape and size control of nanostructures and materials. The results suggest that atomic force microscopy (AFM) would be a powerful measurement tool for satisfying the need for shape and size measurements.

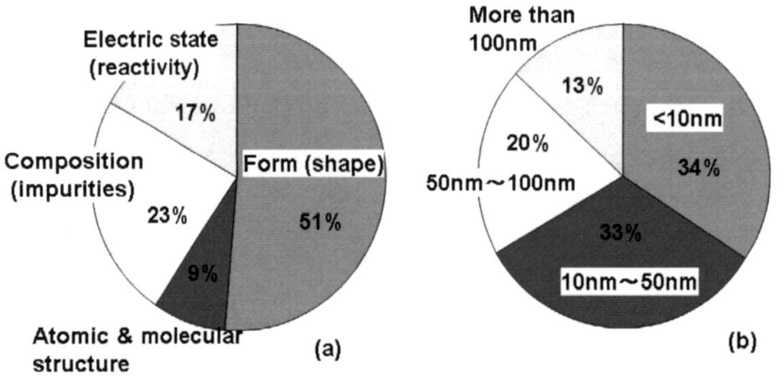

Figure 1. Results of the questionnaire survey on (a) key factor and (b) key size to realize new function by nanotechnology. The analysis bases on the answers from 27 Japanese companies replied to the questionnaire survey. [4].

The standardization of AFM measurements has been discussed by the ISO TC201 (Surface Chemical Analysis) subcommittee (SC) 9 on "Scanning Probe Microscopy." Precise knowledge of AFM probe shapes as well as proper adjustment (calibration) of AFM equipment are essential for accurate nanostructure measurements, since an AFM image is given as the convolution of the actual surface morphology and the probe shape.

The aim of the present paper is to introduce current standardization activities for nanotechnology, especially the measurement and characterization of nanomaterials and nanostructures. Current working items of ISO TC229/WG2 are described in general together with a description of the working item on AFM probe-shape measurement, which is under discussion by ISO TC201/SC9. The reference material prepared for AFM probe-shape measurement is also explained in detail.

2. Standardization activities for nanomaterial measurement and characterization

2.1. Brief overview of ISO TC229 activities

ISO TC229 on "Nanotechnology" began operation in November 2005 and has convened 7 general meetings. The number of participants has increased each time, with more than 200 people attending the 7th meeting held in Shanghai, China in November 2008. At present, 30 national bodies (i.e., countries) serve as P-members (active contributors with the right to vote) and 10 are O-members. The scope of TC229 clearly states that it focuses on standardization in the field of nanotechnologies, which includes the followings:

1) Understanding and control of matter and processes at the nanoscale, typically, but not exclusively, below 100 nanometers in one or more dimensions where the onset of size-dependent phenomena usually enables novel applications.

2) Utilizing the properties of nanoscale materials that differ from the properties of individual atoms, molecules, and bulk matter, to create improved materials, devices, and systems that exploit these new properties.

The present 4 WGs in ISO TC229 are illustrated in figure 2. The specific tasks of each WG are to develop standards for the followings: unambiguous and uniform terminology and nomenclature (WG1); measurement, characterization, and test methods with consideration of metrology and reference materials (WG2); science-based health, safety, and environmental aspects (WG3); and materials specifications (WG4). Among them, WG1 and WG2 are working jointly with IEC TC113 as mentioned above.

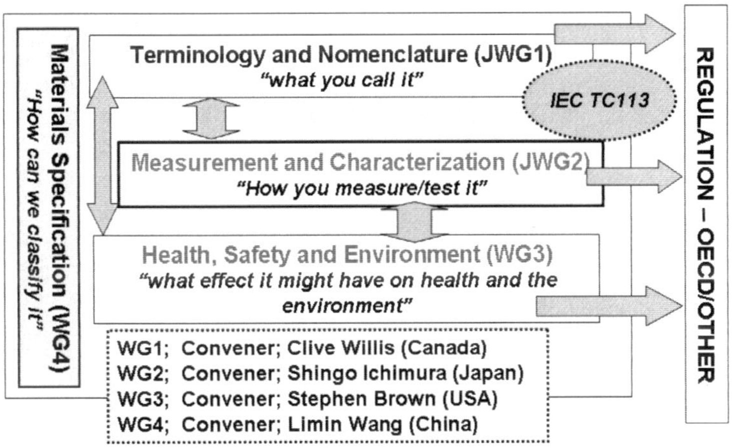

Figure 2. Structure of ISO TC229 on Nanotechnology.

2.2. *Current activities of WG1, WG3, and WG4*

As the output of WG1 and WG3 activities, ISO TC229 has published 2 standardization documents (Table 1): a technical report (TR) and a technical specification (TS).

Table 1. Titles of the standard documents published from ISO TC229

Document No.	WG	Title of document
ISO/TR 12885:2008	WG3	Health and safety practices in occupational settings relevant to nanotechnologies
ISO/TS 27687:2008	WG1	Terminology and definitions for nano-objects - Nanoparticle, nanofibre and nanoplate

The purpose of a TR is to circulate data obtained from surveys carried out among the national bodies or "state of the art" data in relation to standards for a particular subject. A TS, in contrast, usually presents prospective standards for provisional application in reply to an urgent request for guidance to meet an identified need. A TS will usually be reviewed 3 years after its publication to consider its conversion into an International Standard (IS) with additional information.

Since standardization should be aligned with the continuing research and development of nanotechnology, ISO TC229 discusses many more TSs and TRs than ISs. The official time period for the publication of a standardization document is 24 months for a TS or TR, and 36 months for an IS. Other main projects discussed in WG1, WG3, and WG4 include the followings: terminology relating to carbon nano-objects, nanostructured materials, bio-nano interfaces, etc. (WG1); methods for endotoxin and inhalation toxicity testing, and guidelines for the safe handling and disposal of manufactured nanomaterials (WG3); specification guidelines for nano-calcium carbonate, nano-titanium oxide, etc. (WG4). Notably, guidance for the physicochemical characterization of engineered nanoscale materials for toxicological assessment is now under discussion jointly between WG3 and WG2 to help the activity of the Organisation for Economic Co-operation and Development (OECD).

2.3. Current activities of WG2

At the 2nd ISO TC229 meeting held in Tokyo during May 2006, WG2 agreed to set as its first target the standardization of carbon nanomaterials. On the basis of this decision, the convener of WG2 circulated a survey containing a list of various measurement properties and possible measurement methods for experts in member bodies to complete. The survey asked whether each measurement property was important and useful for suppliers or for users with appropriate quality control, and if a given method was already well-established for the characterization of multiwall carbon nanotubes (MWCNTs). The properties and measurement methods that received a high grade of importance in the survey are listed in table 2. The protocols for the measurement of CNTs with the methods, together with their principles and descriptions of experimental results, are now under preparation as TS10929 by the project group (PG5) of TC229 WG2.

Table 2. Category, property, and methods for characterizing MWCNT described in the work item of ISO TC229/ WG2/PG5 (TS10929).

category	property	method
purity	moisture content	weight loss measurement
purity	ash content	weight loss measurement
purity	metallic residual content	ICP-AES[a], XRF[b]
purity	volatile content	weight loss measurement
purity	polyaromatic hydrocarbons	HPLC-MS[c]
purity	carbon materials excluding MWCNT	SEM
purity	disorder	Raman spectroscopy
physical property	burning property	TG/DTA[d]
physical property	stacking nature	XRD[e], TEM
geometrical property	inner diameter	TEM
geometrical property	outer diameter	SEM and/or TEM
geometrical property	length	SEM or TEM
geometrical property	morphology	SEM and/or TEM

[a]ICP-AES: Inductively Coupled Plasma- Atomic Emission Spectroscopy

[b]XRF: X-ray fluorescence analysis

[c]HPLC-MS: High Performance Liquid Chromatograph- mass spectrometer

[d]TG/DTA: Thermogravimetric/ Differeential thermal analysis

[e]XRD: X-ray diffractometer

For the characterization of single-wall carbon nanotubes (SWCNTs), however, the WG2 decided to follow the steps proposed by ANSI (USA), which has proposed the adoption of a layered approach to standardization. The first layer consists of the characterization of purity and structure, which will be followed by the characterization of the electrical, magnetic, mechanical, and optical properties. Moreover, ANSI has proposed to limit the measurement methods for the initial screening step in the first layer. Therefore, the methods chosen for the initial screening step are 1) transmission electron microscopy (TEM), 2) scanning electron microscopy (SEM), 3) ultraviolet (UV)-visible-near-infrared (NIR) absorption, 4) NIR-photoluminescence (NIR-PL) spectroscopy, 5) evolved gas analysis-gas chromatograph mass spectrometry (EGA-GCMS), 6) thermogravimetric analysis (TGA), and 7) Raman spectroscopy. Presently, TSs for the characterization of SWCNTs by these 7 methods are underway in ISO TC229/WG2. Member bodies of ANSI, JISC (Japan), and KATS (Korea) are mainly responsible for the preparation of these documents.

The standardization of CNTs is quite important and urgent since the potential risks of CNTs (e.g., the possibility of inducing mesothelioma) have recently attracted much attention [5, 6]. Experiments for potential risk assessment should use well-established test protocols and well-characterized test samples. Most of the 7 TS documents will be published by the end of 2009, and will provide common protocols for the characterization of test samples. Another discussion item in TC229/WG2 on the "General Framework for Determining Nanoparticle Content in Nanomaterials by Generation of Aerosols," which is led by DIN (Germany), will also contribute to the establishment of test protocols.

3. Standardization activities regarding nanostructure measurements by AFM

3.1. Current problems with AFM measurements

Although AFM is well known to have very high resolving power for spatial imaging, such as in the observation of an atomic arrangement on a solid surface, the following problems have been indicated. When two protrusive nanostructures with a narrow gap between them are measured, the depth measured between the two structures (Hm) is shallower than the height of the protrusion (Ho), as is shown schematically in figure 3(a). In addition, in the measurement of an isolated protrusive nanostructure, the measured width (Wm) is wider than the width of the original structure (Wo), as is shown in figure 3(b). In both cases, the difference between the measured value of the width or height and the real value is related to the shape of the probe used for the AFM observation. The shapes of the probe contour and apex are responsible in the former and latter cases, respectively. Moreover, there is no practical way to know the real probe shape (i.e., probe apex and contour) under in-situ measurement conditions.

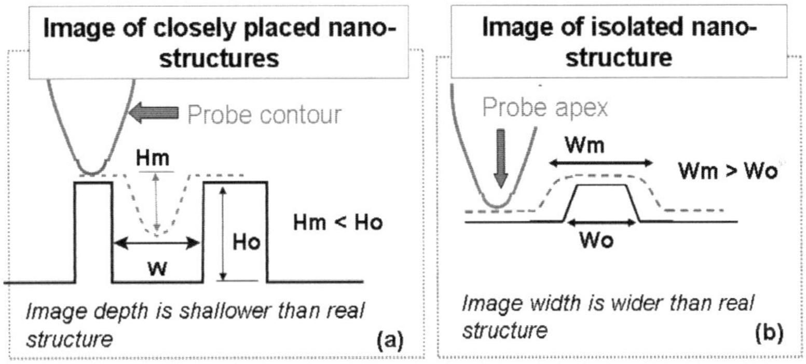

Figure 3. Schematic illustration of AFM measurement results for (a) closely placed two nano-structures and (b) an isolated nano-structure

3.2. AFM probe characterizer for in-situ measurement of a probe shape

To characterize the shape of an AFM probe, especially of the probe contour, it is useful to know the relationship between the probe width (W) and the length measured from the probe apex (L), as is shown schematically in figure 4(a). This relationship can be easily estimated under in-situ conditions when a probe characterizer with a comb-shaped structure (figure 4(b)) is used. The value of W is obtained from the designed gap width of a rectangular hollow in the probe characterizer, and L can be measured from the trajectory (corrugation) of the AFM probe during measurement of the characterizer. The length at a thicker part of the probe (L_2) is obtained from a wider hollow (W_2 in figure 4(b)), while that at a thinner part (L_1) is obtained from a narrower hollow (W_1 in figure 4(b)). Therefore, a probe characterizer with a comb-shaped structure with many different gap widths is necessary to obtain reliable information about the contour of an AFM probe.

As a preliminary step, we fabricated a tip characterizer with a comb-shaped structure having different line and space widths (10, 20, and 60 nm) and a knife-edge structure (figure 5(a)) [7]. For the 20-nm comb-shaped pattern, 10 periodic line/space structures were designed and fabricated considering that it will be possible to calibrate the periodicity of the line/space structures by using metrological electron microscopy or metrological AFM [8]. The calibrated structures will add the function of an absolute length scale to the characterizer. An AFM image of the probe characterizer is shown in figure 5(b). The comb-shaped pattern and the knife-edge structure are clearly imaged, although most of the lines appear wider than their actual sizes owing to the size of the AFM probe used for the observation.

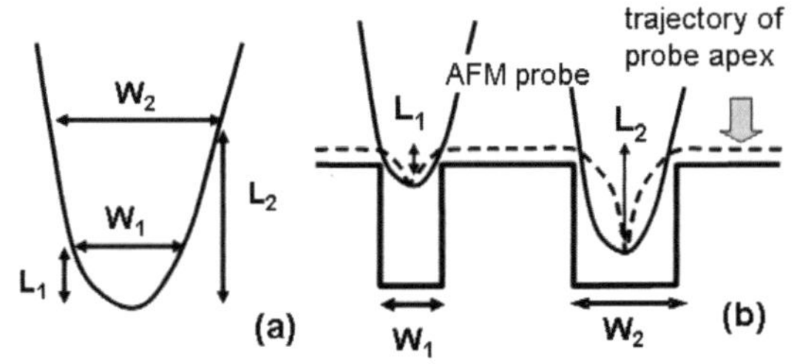

Figure 4. (a)Schema showing the relation of probe width and probe length. (b) An example of comb-shape structure to measure the relation between probe width and probe length.

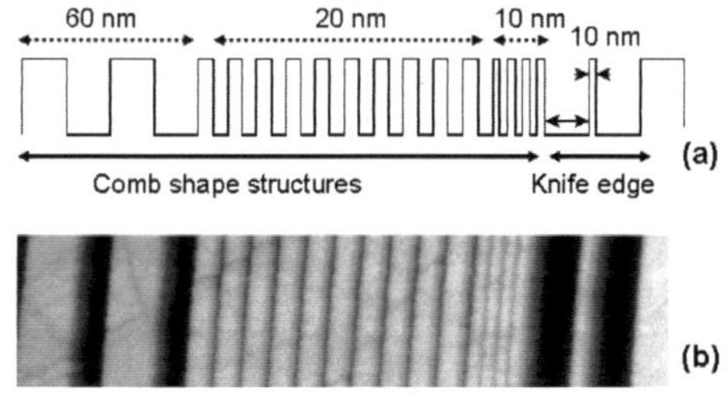

Figure 5. Example of a test probe characterizer with comb-shape structures: (a) design, and (b) an AFM image

The relationship between the probe width and length (i.e., the probe-shape function) can be easily determined by measuring the line profile of the characterizer. Figure 6(a) shows the results of such a line profiling, which yielded corrugation values of 1.6 nm (for the comb-shaped structure with 10-nm width), 4.9 nm (20-nm width), and 25 nm (60-nm width), respectively. Using these data, we obtained the probe-shape function shown in figure 6(b).

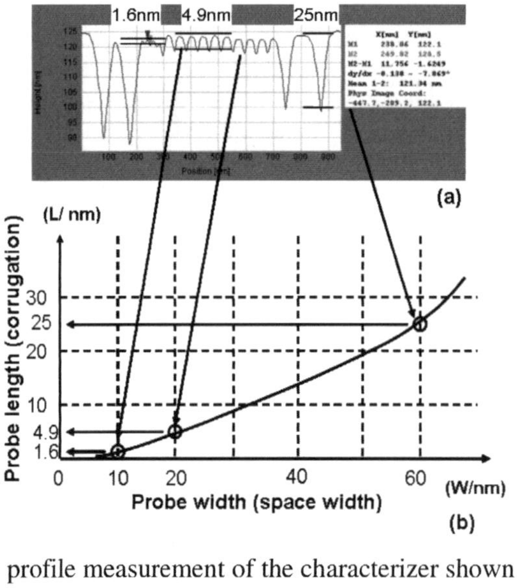

Figure 6. (a)Results of line profile measurement of the characterizer shown in figure 5, and (b) the probe-shape function of an AFM probe used for the line profiling.

3.3. Fabrication of an AFM probe characterizer with a comb-shaped structure

To measure a probe shape precisely, the probe characterizer should have many line and space widths in the range between a few and over a hundred nm. We fabricated another characterizer with 14 line and space structures within the width range from 5 to 100 nm [9]. It is well known that the metal organic chemical vapor deposition (MOCVD) method allows control of film thickness with high accuracy for a semiconductor compound such as GaAs. Thus we have adopted the MOCVD method for the fabrication of the probe characterizer. Details of the fabrication process are described elsewhere [7]; here we briefly introduce the process for better understanding of the probe characterizer.

GaAs and GaInP layers were alternately grown by the MOCVD method on a GaAs substrate to form multiple sets of GaAs/GaInP layers. The thicknesses of the layers were controlled precisely to realize the desired structure. The thickness accuracy depended on the deposition rate and the intermixing of layers at the interfaces: the error was less than 2 nm in the case of the characterizer shown in figure 5. After the film was grown on a GaAs wafer by the MOCVD method, the wafer was cut into pieces in the vertical direction. The cross-sectional surface of each piece was polished until the surface roughness was less than 1 nm rms (root mean square), and then was etched with a solution containing sulfuric acid and hydrogen peroxide. Since GaAs has an etching rate in this solution about 1000 times higher than GaInP, only the GaAs parts were selectively etched while the GaInP parts retained their original height. Thus the probe characterizer could be fabricated with GaInP layers as lines and GaAs layers as spaces in a comb-shaped structure.

3.4. Standardization and utility of the in-situ probe shape measurement with the characterizer

A new work item with the title "Procedure for in-situ characterization of AFM probes used for nanostructure measurement" has been submitted to ISO TC201/SC9. It is clearly stated in the work item that the standard method is appropriate for characterizing the profile of a probe by determining a probe-shape function, and that it can reduce the uncertainty of AFM measurements of nanostructures.

Nanostructures are now widely used in practical industrial applications. For example, the sensor of a magnetic hard disk is designed to have a sharp concave structure (38 nm in depth and 65 nm in width). However, because it is still premature to use AFM observation for industrial applications and standardization is needed for reliable measurements, cross-sectional TEM observation has been used to evaluate fabricated nanostructures although its cost is higher and the sample preparation time is longer [10].

If we want to measure a fabricated structure by AFM, we have to determine whether the probe to be used for the measurement is sharp enough. A measurement of the probe-shape function based on the proposed standard will make the determination, as shown schematically in figure 7(b). The green area corresponds to the condition that the probe length is larger than 38 nm when its width is narrower than 65 nm. Since the probe-shape function of probe A passes through the green area, but that of probe B does not, only probe A can be used for the AFM measurement.

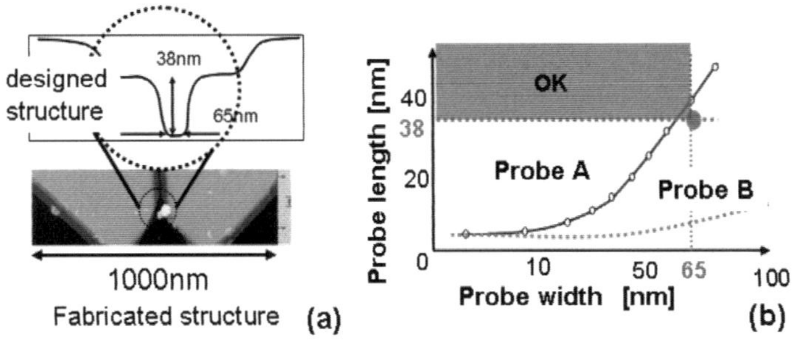

Figure 7. (a) Example of an industrial application of nanostructure. (b) Example for the judgment of AFM probe applicability using a probe-shape function.

Another important issue for the measurement of the probe-shape function by means of the characterizer is that we have to control and adjust the amplitude of the cantilever used for the AFM observation. To optimize the motion of the AFM cantilever, the standard protocol recommends measuring the knife-edge structure together with the comb-shaped structures. The details will be addressed, and the results will be published in the near future. An accurate probe-shape function can also be used to evaluate the true dimensions of nanomaterials by means of the reproducible restoration technique for distorted AFM images [11].

4. Summary

The current standardization activities on nanotechnology, especially those related to the measurement and characterization of nanomaterials and structures and that have been promoted by ISO TC229/WG2 and ISO TC201/SC9, have been described in detail. The standardization activities not only assist industrial applications of nanomaterials and structures, but also support assessment of their potential risks to human health.

TC229/WG2 started its activity by focusing on carbon nanomaterials, especially CNTs. Other target items of standardization will be the measurement and characterization of fullerene together with those for engineered nanoparticles such as titanium dioxide, zinc oxide, and silicon dioxide. These materials are targeted because they are produced in larger amounts than SWCNTs and MWCNTs, and because they have been selected as target materials by OCED working programs. The standardization of coatings and nanostructured materials (composites and porous structures) will also be discussed together with the basic metrology at the nanoscale as described in the outline strategy for ISO TC229/WG2 [12].

Acknowledgements

The authors would like to express their sincere appreciation to Dr. K. Yanagiuchi of TDK Corporation for the discussion on the industrial application of AFM measurements. Thanks are also due to Ms C. Wang for her assistance in the AFM observations of the characterizer. This work was partially supported by the international joint research program (NEDO Grant) for International Standard Development Area (2005IS128).

References

[1] Roco M 2006 "Nanotechnology R&D in the Americas and the Global Context", in *2nd International Dialogue on Responsible Research and Development of Nanotechnology* (Tokyo/Japan, June 27-28, 2006)

[2] Peng G S 2006 "Nanometrology Research and Its Standardization in Taiwan", in the *Proceedings of the 2nd Intern. Symp. on Standard Materials and Metorology* (Tokyo/Japan, June 25-26, 2006) p167

[3] See, for example, Proffitt F 2004 *Science* **305** 762, Maynard A D 2006 *Nature* **444** 267

[4] The survey was performed and reported in 2005 by the committee of investigation on nanotechnology standardization organized by the Japan Standard Association.

[5] Takagi A, Hirose A, Nishimura T, Fukumori N, Ogata A, Ohashi N, Kitajima S, and Kanno J 2008 *J. Toxicol. Sci.* **33** 105

[6] Poland C A , Duffin R, Kinolch I, Maynard A, Wallace W A H, Seaton A, Brown V S, MacNee W, and Donaldson K 2008, *Nature Nanotechnology* **3**, 423

[7] Itoh H, Fujimoto T, and Ichimura S 2006, *Rev. Sci. Instrum.* **77** 103704

[8] Misumi I, Gonda S, Sato O, Sugawara K, Yoshizaki K, Kurosawa T, and Takatsuji T 2006, *Meas. Sc. Technol.* **17** 2041

[9] Wang C, Itoh H, Homma Y, Sun J, Hu J, and Ichimura S 2008, *Jpn J. Appl. Phys.* **47** 6128

[10] Yanagiuchi K (private communication)

[11] Fujita D and Xu M 2008, in the abstract of the IUMRS International Conference in Asia 2008 (Nagoya/Japan, December 9-13, 2008) NI-2

[12] *"Outline strategy for ISO TC229/ WG2 Ver. 8.0"* 2008 (unpublished), prepared by the Study Group on Strategy of ISO TC229/ WG2 (led by Hossain K/ NPL)

Standardization of Nanomaterials Characterization by Scanning Probe Microscopy for Societal Acceptance

Daisuke Fujita[1], Keiko Onishi[2] and Mingsheng Xu[3]

[1] International Center for Materials Nanoarchitectonics (MANA) and Advanced Nano Characterization Center (ANCC), National Institute for Materials Science (NIMS), 1-2-1 Sengen, Tsukuba, Ibaraki 305-0047, Japan

[2] Advanced Nano Characterization Center (ANCC), National Institute for Materials Science (NIMS), 1-2-1 Sengen, Tsukuba, Ibaraki 305-0047, Japan

[3] International Center for Young Scientists-Interdisciplinary Materials Research (ICYS-IMAT), National Institute for Materials Science (NIMS), 1-2-1 Sengen, Tsukuba, Ibaraki 305-0047, Japan

E-mail: fujita.daisuke@nims.go.jp

Abstract. Novel nanomaterials are expected to play key roles for the promotion of innovations in the various industrial products. In order to make such novel nanomaterials to be socially acceptable and widely used, it is very important and necessary to establish the reliable nano-characterization methodology for the industrial nanomaterials under the authorized international scheme for standardization. Among the nano-characterization methods, scanning probe microscopy (SPM) is the most versatile both in the measurement functions and the operational environments. Whereas there are various nanomaterials of industrial application, fullerene nanomaterials (FNM) have attracted much attention due to their unique physical properties. Here we show the importance of the quantitative analysis and standardization of SPM using FNM as a typical example.

1. Introduction

Nanomaterials can be defined as engineered structures with at least one dimension of approximately 100 nanometers or less [1]. Since the functional nanomaterials are expected to play key roles for the promotion of innovations in the various industrial fields, these nanomaterials have been increasingly employed for the commercial purposes during this decade. With the increase of the engineered nanomaterials circulated in the society, the potential impact and risk on the environment and the health of human being have attracted a growing attention in the world [2]. Especially the potential cytotoxicity of nanoscale carbon materials such as single-wall nanotube (SWNT), multi-wall nanotube (MWNT), and fullerene-based nanomaterials has been a matter of interests among the so-called nano-toxicologists [3,4]. There has been a problem in the quantitative evaluation of nanoscale toxicity and conflicting results on the biological effects of carbon nanomaterials have been reported, in which some reports high and others low toxicity. The observed discrepancy is mainly caused by the distribution of size and the impurity concentration such as the metal catalyst and the surfactants used for solubilization. Therefore, the establishment of quantitative characterization methodology at the nanoscale is now considered to be pre-requisite for the reliable risk assessment.

On the contrary, in order to trade the engineered nanomaterials globally, it is also required to specify the properties of the nanomaterials using the unified terminology, nomenclatures, definitions, and measurement methodology. Especially, how to measure the basic properties at the nanoscale, such as dimensions, size distribution, atomic structure, chirality, impurity content, electrical conductivity, magnetic property, specific surface area, and so on, shall be important items to be specified in a unified manner. Thus, in order to make the novel nanomaterials be socially acceptable and be smoothly traded in the world, it is important to establish the reliable nanocharacterization methodology under the authorized international scheme for standardization [5,6].

Among the nanoscale characterization methods, scanning probe microscopy (SPM) can probe the most powerful and versatile method. Top two of the various important innovations in the SPM development are the inventions of scanning tunneling microscopy (STM) and atomic force microscopy (AFM), which were invented in 1982 and 1986, respectively [7,8]. Since then, various kinds of SPM have been developed, which enables us to probe a variety of physical and chemical quantities. Owing to the compactness, SPM has a high environmental adaptability. SPM can be operated not only in air but also in liquid and in ultrahigh vacuum. Furthermore, SPM probes can be used as versatile tools for nanoscale fabrication and atom manipulation [9]. Nowadays, SPM is used not only for basic research but also for fabrication and inspection processes in industry.

In this perspective article, the recent activities and researches related to the standardization of nanotechnologies and SPM standardization is described firstly. These activities are performed in two technical committees (TC229 and TC201) of the International Organization for Standardization (ISO). Next, current challenges in the SPM standardization on quantitative analyses are discussed based on the ISO/TC201 activities [10]. Besides the quantitative reduction methodology of morphological artifacts caused by the finite size of SPM tips is introduced [11,12]. Finally, actual applications of the AFM image restoration methodology to the actual nanomaterials are explained.

Here, among various nano objects, fullerene nanomaterials (FNM) are focused since they have attracted much attention due to their unique physical properties. Especially low-dimensional fullerene nano-crystals such as fullerene nanowhiskers (FNW) and nanotubes (FNT) are attractive since they exhibit semiconducting transport property. Actually the electric property of FNWs has been clarified by using a conductive AFM method [13]. Such low-dimensional semiconductor FNMs may represent promising building blocks for the bottom-up assembly of integrated electronic and photonic systems. Obviously the most significant role of AFM is the precise evaluation of the dimensions of nanostructures. Effectiveness of the image restoration methodology for distorted AFM images using an actual tip shape function is demonstrated.

2. International Standardization in Nanotechnologies

A brief history of the international standardization in the area of nanotechnology is schematically shown in Fig.1. The first initiative for pre-standardization research on nanotechnologies within an international collaboration scheme was thought to be the establishment of a new technical working area (TWA29) on "materials properties at the nanoscale" of VAMAS (Versailles Project on Advanced Materials and Standards) in 2002. VAMAS supports world trade in products dependent on advanced materials technologies, through international collaborative projects aimed at providing the technical basis for harmonized measurements, testing, specifications, and standards [14]. Recently VAMAS has approved to establish another new technical working area (TWA34) on "nanoparticle populations", which will first deal with refining the measurement of chirality distribution in polydisperse SWNT sample. The main mission of VAMAS activity is to promote the international collaborative research on pre-standardization through the international round robin tests. The final technical reports shall be transferred to the corresponding ISO or IEC (International Electrotechnical Commission) committees for the detailed refinement towards international standards.

Actual standardization initiative in comprehensive nanotechnologies was first started in Europe, where CEN (European Committee for Standardization) established BTWG-166 on nanotechnology in 2004. CEN/BTWG-166 has been succeeded by the establishment of CEN/TC352 on "Nanotechnologies" at the end of 2005. During the same period, the American National Standard Institute (ANSI) launched Nanotechnology Standard Panel (ANSI-NSP) in 2004, serving as the cross-sector coordinating body for the purposes of facilitating the development of standards in the area of nanotechnology. Soon after ASTM International Committee E56 (ASTM-E56) on nanotechnology was formed in 2005. In Japan, Japanese Standard Association (JSA) formed Nanotechnology Standardization Panel in 2004, which was followed by the establishment of Japanese Nanotechnology Standardization Committee by JISC (Japan Industry Standards Committee) in 2005.

The proposal of a new technical committee (TC) dealing with the standardization of nanotechnologies on an extensive scale was done by British Standard Institution (BSI) in 2004 within the framework of ISO, which resulted in the establishment of ISO/TC 229 on "Nanotechnologies" in 2005. A year later, another technical committee dealing with electrical aspect of nanotechnology was set up in IEC as TC113. The purpose of the IEC TC113 is to deal with the relevant nanotechnological aspects in developing generic standards for electrical products and systems. Among the above activities of standardization in nanotechnologies, the scheme of ISO/TC229 is thought to be the most world-wide, since it has made liaisons with the related activities for standardization or pre-standardization of nanotechnology. TC229 has first established three working groups (WGs) such as JWG1 (terminology and nomenclature), JWG2 (measurement and characterization) and WG3 (health, safety and environment), in which two are joint working groups with IEC TC113. As for the measurement and characterization, the standardization of measurement methodologies for carbon nanomaterials such as SWNT, MWNT and fullerene have been selected as highly prioritized working items in JWG2.

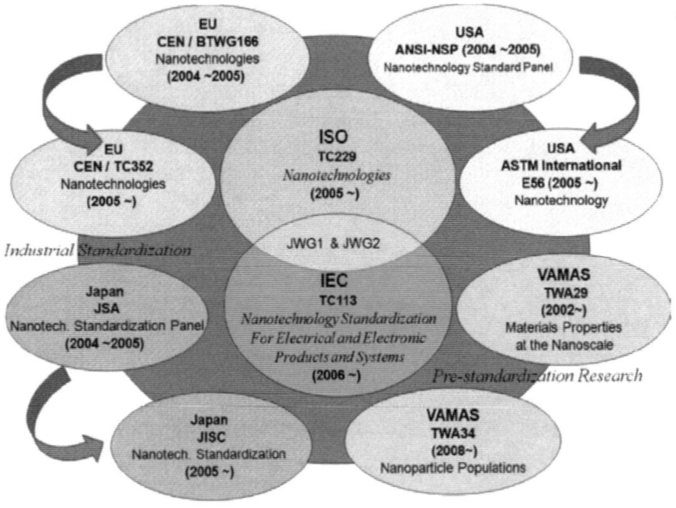

Fig.1. A rough view of international standardization activities related to nanotechnology.

3. International Standardization of Scanning Probe Microscopy

The standardization activities on SPM were formed in the ISO framework in 2004 when Subcommittee 9 for SPM was established in TC201 for surface chemical analysis [10]. The most prioritized task was to standardize the terminology of SPM, because it will provide a firm basis for the following standardizations. It includes acronyms for SPM-specific terms, definitions and terms for

SPM methods, and acronyms and terms for contact mechanics. The following standardization on SPM shall proceed towards the standardization of data management and treatment, dimensional calibration, reference materials, and guidelines for operational procedures. The priority of the standardization is generally based on the actual needs of SPM users, manufacturers and vendors. Since SPM has a large versatility in functions and applications, the standardization should deal with not only quantification of dimensional analyses by topographic imaging but also atomic-scale chemical analyses, fabrication and manipulation techniques, electric and magnetic property analysis, nanomechanical property analysis and so on. Moreover, guidelines for the proper uses of SPMs in different environments such as in air, in liquids, in UHV, and so on, should be standardized.

The major functions of SPM as analytical methods are imaging and spectroscopy at the nanoscale. The acquired raw data may require appropriate data managements and treatments. While a wide variety of SPM instruments have been manufactured since the invention of STM, a wide variety of data formats using different terminologies have been created by the SPM manufacturers. This situation may cause inconveniences for the SPM users to exchange data collected with different SPM instruments. The development of a standard data format shall facilitate inter-exchangeability and consistent treatment of SPM data, enhance the productivity of data processing programs, and increase accuracy and quantification. The proposed format consists of an information header and subsequent data array. Since the current data transfer format for surface chemical analysis (ISO 14976) is composed of text [15], the proposed data format for SPM is written in text format. The obvious advantage is that the data file can be accessed, independent of computer system, using text editor. The proposed arrangement of data array and the examples SPM data taken by different SPM instruments are shown in Fig.2. Once transferred to the standard format, it is easy to compare the individual data using the unified treatments.

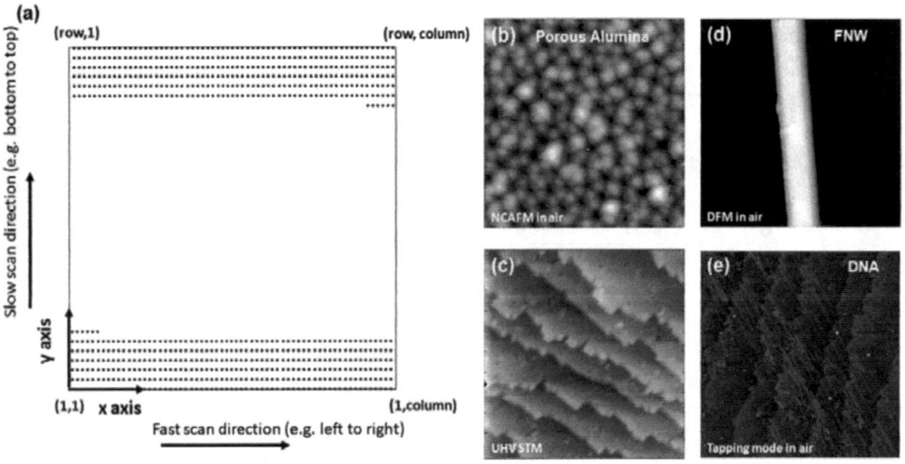

Fig.2. (a) Geometry and arrangement for a two-dimensional map data array. (b)-(e) SPM images taken by different instruments and expressed by the data transfer format.

4. Quantitative Morphology Imaging of Nanomaterials

For quantitative morphology imaging of nanomaterials, it should be noted that significant distortion in imaging may occur if the sample surface has large corrugation compared to the tip apex shape [16]. The real surface topography, s(x,y), is dilated to the image surface topography, z(x,y), by the finite tip shape, t(x,y):

$$z(x, y) = \max\{s(x', y') - t(x'-x, y'-y)\} \tag{1}$$

The dilated topography image can be partially restored by using the erosion algorithm to reconstruct the upper bound image, r(x,y), of the real surface topography:

$$r(x,y) = \min\{z(x',y') + t(x'-x, y'-y)\} \tag{2}$$

A numerical simulation of scanning (dilation) and reconstruction (erosion) of AFM topography imaging using a known tip-shape function is shown in Fig.3. Since there is a great need for a reproducible procedure for restoring AFM images, one should be standardized by the ISO.

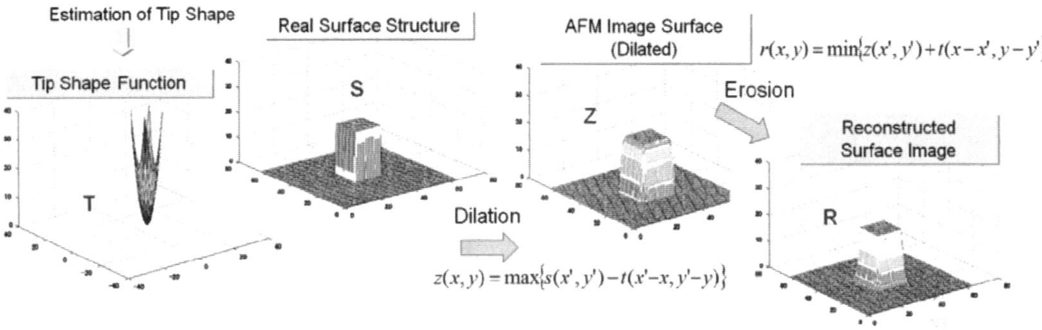

Fig.3. Deducing reconstructed surface r(x,y) from AFM-scanned surface and a tip shape function.

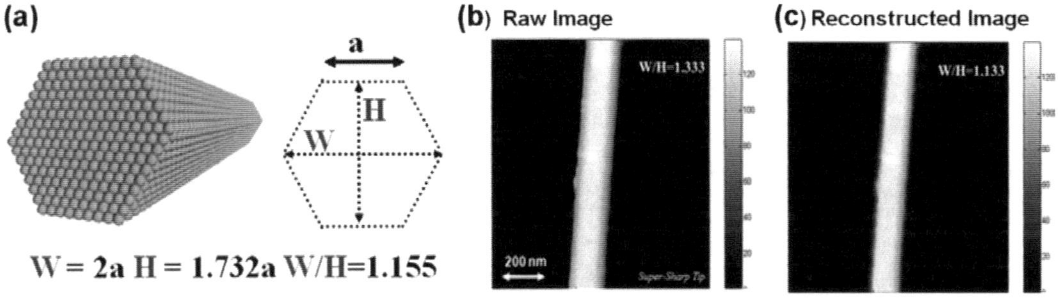

Fig.4. Structure model of FNW (a), a raw AFM topography image of a FNW (b), and its reconstructed image using a tip shape function (c).

For the engineered nanomaterials, one of the first prioritized properties to be specified quantitatively is their dimensions such as length, height, width, diameter, and so on. In order to realize the nanomaterials to be accepted as industrial products by market, it is necessary to establish quantitative methodology for critical dimension (CD) measurements using a noncontact mode AFM topography imaging. As for one-dimensional nano-objects such as FNWs, precise CD measurements of height, length and width are required, if they are to be traded in market. Typical crystal structure and outer shape of FNW is shown in Fig.4(a), which is likely to have a hexagonal cross-section. Compared with the raw AFM image shown in Fig.4(b), the reconstructed image using the tip shape function shows a significant reduction in width and sharpens the image as shown in Fig.4(c). The observed ratio of width/diameter is found to be consistent with the typical model structure shown in Fig.4 (a). However, it should be noted that the morphology of individual FNWs can be different.

In order to apply the image reconstruction in a quantitative and reproducible manner, it is critical to evaluate the tip shape function in use as precisely as possible using novel tip characterizers [17]. Since the quantitative CD measurement procedure of artificial nanoparticles at about 1 nm resolution should have potential industrial demands, pre-standardization research shall be performed.

5. Conclusion

In this perspective paper, we described the importance of international standardization of nanotechnologies and nanomaterials in order to make nanotechnology products be acceptable to the society and market. We pointed out that the precise and quantitative characterization of nanomaterials is prerequisite not only for fair evaluation of potential risk or toxicity of novel nanomaterials, but also for global trading of engineered nanomaterials or nano-products. Based on the strong demands for the nanotechnology standardization and pre-standardization research, various activities have been initiated in EU, US and Japan since 2002. After the establishment of ISO/TC229 on nanotechnologies, global standardization of nanotechnologies can be performed and discussed intensively since 2005.

Here we show the importance of the quantitative analysis and standardization of SPM using FNM as a typical example. Restoration of a dilated AFM topography image of FNW is successfully demonstrated. Since SPM can be used for the measurements of various materials properties, we have to promote the pre-standardization research for the versatility of SPM nanocharacterization.

Acknowledgement

The authors would like to appreciate Dr. Miyazawa for the kind offer of the FNW samples. This work was partially supported by Japan Science and Technology Agency and by New Energy and Industrial Technology Development Organization (NEDO) - international joint research grant program. This work was also in part supported by World Premier International Research Center Initiative (WPI Initiative) on Materials Nanoarchitectonics, MEXT, Japan.

References

[1] Nel A, Xia T, Madler L and Li N 2006 *Science* **311** 622
[2] Colvin V L 2003 *Nature Biotechnology* **21** 1166
[3] Jia G ,Wang H, Yan L, Wang X, Pei R, Yan T, Zhao Y and Guo X 2005 *Environ. Sci. Technol.* **39** 1378
[4] Isobe H, Tanaka T, Maeda R, Noiri E, Solin N, Yudasaka M, Iijima S and Nakamura E 2006 *Angew. Chem. Int. Ed.***45** 6676
[5] Todua P A 2008 *Measurement Techniques* **51** 462
[6] Rashba E and Gamota D 2003 *J. Nanopart. Res.* **5** 401
[7] Binnig G and Rohrer H 1982 *Helv. Phys. Acta.* **55** 726
[8] Binnig G, Quate C F and Gerber Ch 1986 *Phys. Rev. Lett.* **56** 930
[9] Fujita D and Sagisaka K 2008 *Sci. Technol. Adv. Mater.* **9** 013003
[10] Fujita D, Itoh H, Ichimura S and Kurosawa T 2007 *Nanotechnology* **18** 084002
[11] Onishi K and Fujita D 2008 *J.Vac. Soc.Jpn.* **51** 165
[12] Xu M, Fujita D, Onishi K and Miyazawa K 2009 *J. Nanosci. Nanotechnol.* **9** in press.
[13] Xu M, Pathak Y, Fujita D, Ringor C and Miyazawa K 2008 *Nanotechnology* **19** 075712
[14] http://www.vamas.org/
[15] ISO 14976: 2000 *Surface chemical analysis – Data Transfer Format*
[16] Keller D 1991 *Surf. Sci.* **253** 353
[17] Itoh H, Fujimoto T and Ichimura S 2006 *Rev. Sci. Instrum.* **77** 103704

Biological Activities of Water-soluble Fullerene Derivatives

S Nakamura and T Mashino

Department of Pharmaceutical Sciences, Faculty of Pharmacy, Keio University, 1-5-30 Shiba-koen, Minato-ku, Tokyo 105-8512, Japan

E-mail: mashino-td@pha.keio.ac.jp

Abstract. Three types of water-soluble fullerene derivatives were synthesized and their biological activities were investigated. C_{60}-dimalonic acid, an anionic fullerene derivative, showed antioxidant activity such as quenching of superoxide and relief from growth inhibition of *E. coli* by paraquat. C_{60}-bis(*N,N*-dimethylpyrrolidinium iodide), a cationic fullerene derivative, has antibacterial activity and antiproliferative effect on cancer cell lines. The mechanism is suggested to be respiratory chain inhibition by reactive oxygen species produced by the cationic fullerene derivative. Proline-type fullerene derivatives showed strong inhibition activities on HIV-reverse transcriptase. The IC_{50} values were remarkably lower than nevirapine, a clinically used anti-HIV drug. Fullerene derivatives have a big potential for a new type of lead compound to be used as medicine.

1. Introduction

Recently, fullerenes have gained attention in relation to toxicities of nano-materials. Suspension of fullerene was reported to increase oxidative stress in fish brain [1] or embrio [2]. On the other hand, problems of residual organic solvent for preparation of suspension are noted [3].

Fullerene itself is able to behave as a nano-material by aggregating in aqueous solution. However, the most different characteristic of fullerene from the other nano-materials, e.g. carbon nanotube and asbestos, is that fullerene is a molecule with the size of 7 to 8 Å in spherical shape. This means that fullerene is not a nano-material when it behaves as a molecule. Water-soluble fullerene derivatives should act differently as a molecule from unmodified fullerene.

Biological activities of water-soluble fullerene derivatives have attracted many researchers' attention. Generally, biological activities of a compound are related to its chemical properties. Some biological effects based on chemical reactivities of fullerenes have been reported. Fullerene derivatives with polyethylene glycol moiety were designed as an agent for photodynamic therapy [4] because fullerene is easily excited by light and produces singlet oxygen ($^1O_2$). Gadolinium-encapsulated fullerene is hopeful for use as an MRI-imaging agent [5]. The C_{60} core was reported to fit the hydrophobic substrate binding site of HIV protease, and the fullerene derivatives inhibit this enzyme [6].

The present problems of the medicinal chemistry are, first, that there is almost no medicine for intractable diseases, such as neurodegenerative diseases (Alzheimer and Parkinson's disease), cancer, and virus infection (AIDS, hepatitis C type, and influenza); and secondly, drug tolerance. To overcome these problems, the development of new ideal and epoch-making medicine is desired. The requirements of such medicine are novel chemical skeleton and novel chemical and physical properties. We believe that fullerene has these requirements, and we intend to develop fullerene derivatives as a new type of lead compound to be used as medicine.

We have synthesized some types of water-soluble fullerene derivatives and have reported their biological effects. This paper describes antioxidant, antibacterial, and antiproliferative activities of water-soluble fullerene derivatives. Inhibition activities on HIV reverse transcriptase (HIV-RT) of amino acid-type fullerene derivatives are also investigated.

2. Antioxidant activity of anionic fullerene derivatives

Reactive oxygen species (ROS) such as superoxide anion radical ($O_2^{\bullet-}$), hydroperoxide (ROOH), and hydroxyl radical (•OH) produced during oxidative stress cause various kinds of biological damage. Redox-active compounds often affect the production and/or decomposition of ROS in biological systems. Fullerene is a redox-active compound since it has a low LUMO level and a high HOMO level.

A polyhydroxylated C_{60} derivative (fullerenol) was reported to have quenching activity for $O_2^{\bullet-}$ [7]. However, the properties of fullerenol are quite different from those of the parent C_{60}, since conjugated double bonds are widely broken. Therefore C_{60}-dimalonic acid ($C_{62}(COOH)_4$, **1**, Fig. 1) was selected, because it has sufficient water solubility and is thought to have rather similar properties to C_{60}. $C_{62}(COOEt)_4$ was synthesized and its equatorial isomer was purified and then converted to **1**. $O_2^{\bullet-}$ quenching activity of **1** was examined in the xanthine/xanthine oxidase system. The production of $O_2^{\bullet-}$ was measured in terms of cyt. *c* reduction determined from the increase in absorbance at 550 nm.

Figure 2 shows the effect of **1** on cyt. *c* reduction in the xanthine/xanthine oxidase system. When xanthine oxidase was added to the solution to generate $O_2^{\bullet-}$, cyt. *c* was reduced and the absorbance at 550 nm increased. The increase stopped at around 10 minutes after the addition of xanthine oxidase because cyt. *c* was completely reduced. When **1** was added to the complete system (B in Fig. 2), the initial rate of increase of the absorbance at 550 nm was slowed down. These results indicated that $O_2^{\bullet-}$ was quenched by **1**.

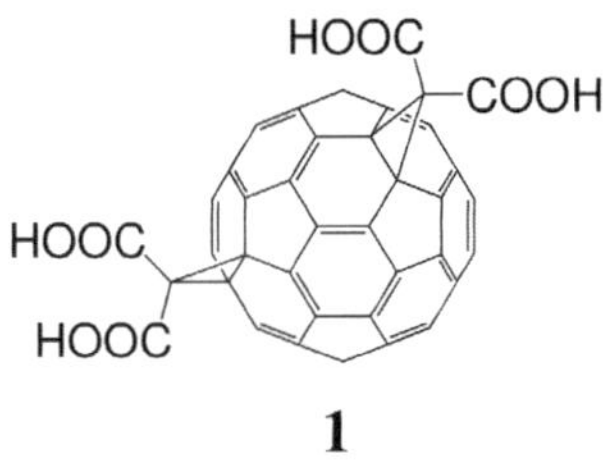

1

Figure 1. Structure of C_{60}-dimalonic acid, **1**.

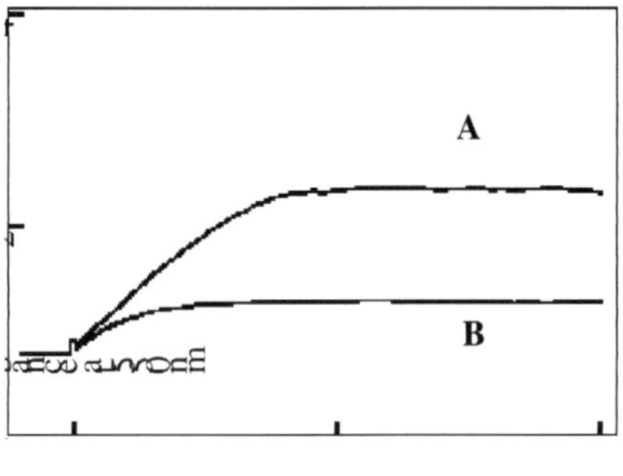

Figure 2. The effect of **1** on cyt. *c* reduction by xanthine and xanthine oxidase.

A. The complete system contained cyt. *c* (10 μM) and xanthine (50 μM) in buffer solution (50 mM phosphate buffer pH 7.8) at 25 °C.

B. **A** + **1** (50 μM)

Paraquat is a typical $O_2^{\bullet-}$-generator that is easily reduced by biological systems such as the respiratory chain and the one-electron-reduced form is able to pass an electron to O_2 to produce $O_2^{\bullet-}$. Paraquat inhibited the growth of *E. coli* by the toxicity of $O_2^{\bullet-}$ (Fig. 3, b). The growth inhibition was relieved by the addition of **1** by quenching of $O_2^{\bullet-}$ (Fig. 3, c – f).

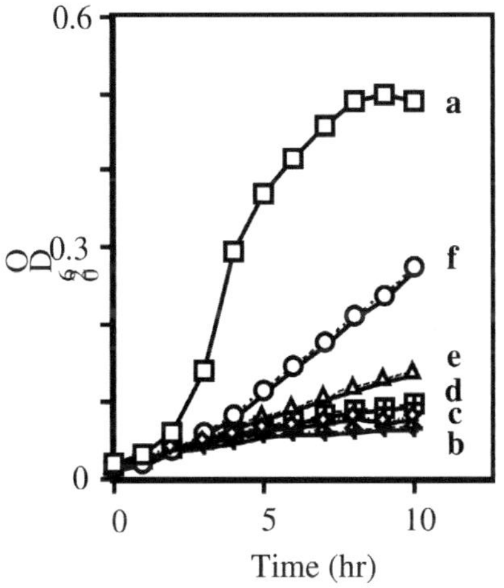

Figure 3. Effect of **1** on *E. coli* growth in the presence of paraquat, an active oxygen generator.
a: control; **b**: paraquat (0.5 μM); **c**: paraquat (0.5 μM) + **1** (10 μM); **d**: paraquat (0.5 μM) + **1** (25 μM); **e**: paraquat (0.5 μM) + **1** (50 μM); **f**: paraquat (0.5 μM) + **1** (100 μM)

3. Antibacterial activities of cationic fullerene derivatives

In the previous section, we demonstrated that **1** decreased the toxicity of reactive oxygen generators in *E. coli*. During the same experiment with other fullerene derivatives, we found the bacteriostatic effect of C_{60}-bis(*N,N*-dimethylpyrrolidinium iodide) (Fig. 4, **2**). Therefore, the inhibition of *E. coli* growth by cationic fullerene derivatives and the inhibition mechanism were investigated.

Cationic fullerene derivative **2** was used as a mixture of regioisomers. Figure 5 shows the effects of fullerene derivatives on *E. coli* growth. Anionic fullerene derivative **1** did not affect growth up to 50 μM and the growth was slightly suppressed at 100 μM (Fig. 5A). On the other hand, only 5 μM of **2** inhibited the growth completely (Fig. 5B).

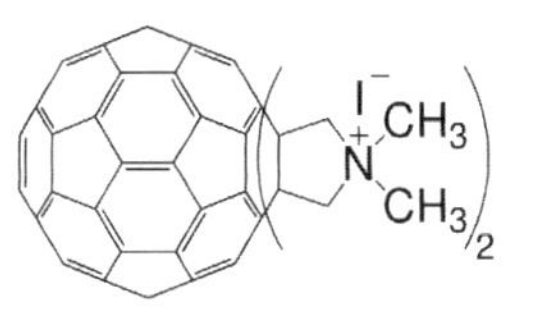

2

Figure 4. Structure of C_{60}-bis(*N,N*-dimethylpyrrolidinium iodide), **2**.

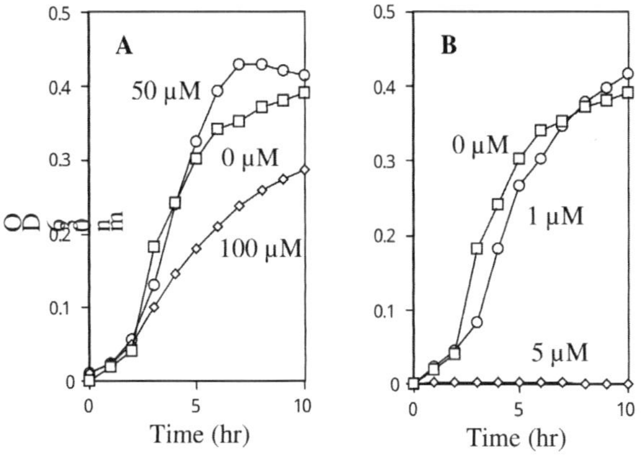

Figure 5. Effect of fullerene derivatives on *E. coli* growth.
E. coli was cultured at 37 °C in a water bath shaker at 100 rpm. Fullerene derivatives were dissolved in DMSO and then added to the medium (A: fullerene derivative **1**, B: fullerene derivative **2**).

To elucidate a mechanism of growth inhibition, O_2 uptake by *E. coli* was examined in the presence of glucose. These fullerene derivatives seemed to inhibit energy metabolism in *E. coli*. Glucose was metabolized through a glycolytic pathway, TCA cycle, and then the respiratory chain, and finally O_2 was consumed to produce H_2O.

As shown in Fig. 6A, the addition of an *E. coli* solution initiated O_2 consumption in the presence of glucose. Dioxygen uptake rates were measured and are shown as a dashed line in Fig. 6. In the presence of 50 μM of **2**, the initial O_2 uptake rate was almost the same as that of the rate without **2** (46.6 μM/min), but after a short lag time (less than 30 seconds), consumption was inhibited (Fig. 6B, 4.1 μM/min). The addition of 50 μM of **1** did not affect consumption (42.2 μM/min).

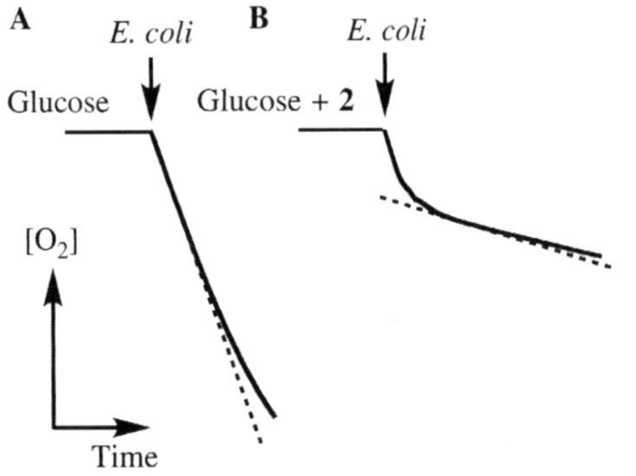

Figure 6. Dioxygen consumption by *E. coli*.
E. coli was grown in a 60-mL medium for 18 hr at 37 °C and then collected by centrifugation. The consumption of O_2 was monitored polarographically with a Clark type electrode. The reaction was started by the addition of 40 mL *E. coli* solution at 37 °C. Fullerene derivatives were also dissolved in DMSO and then added to the reaction buffer.

The result of O_2 consumption inhibition was consistent with that of bacteriostatic effect. That is, **2** was effective in both the bacteriostatic effect and the O_2 uptake inhibition (inhibition of energy metabolism). Finally, **1** had almost no activity. These results strongly indicate that the mechanism of the bacteriostatic effect was energy metabolism inhibition in *E. coli*. As a result, cationic fullerene derivative was suggested to inhibit the respiratory chain from the effect of **2** on O_2 uptake caused by the inner membrane and NADH. These data mean that **2** was reduced by biological reductant system to anion radical of C_{60} and this species reduced O_2 to produce $O_2^{\bullet-}$, in contrast to the $O_2^{\bullet-}$-quenching activity of **1** described in the previous section. The reduction potentials of **1** and **2** were –670 and –480 mV (*vs.* SCE), respectively. This means the reduction of cationic fullerene derivative **2** were easier than that of the anionic derivative **1**. An electrochemical study revealed that reduced fullerene derivative **2** reacts with O_2 and produces H_2O_2 *via* $O_2^{\bullet-}$ [8].

These results prompted us to investigate the antibacterial activities of the fullerene derivatives. Therefore, antibacterial activities of C_{60}-bis(*N,N*-dimethylpyrrolidinium iodide) regioisomers and alkylated C_{60}-bis(*N,N*-dimethylpyrrolidinium iodide) derivatives were investigated (Fig. 7).

Antibacterial activity of the fullerene derivatives *in vitro* was tested against Gram-positive bacteria, according to NCCLS guidelines as minimal inhibitory concentration (MIC) (Table 1). The fullerene derivative was added as a DMSO solution.

The regioisomers of C_{60}-bis(*N,N*-dimethylpyrrolidinium iodide), **2-*t*-2**, **2-*t*-3**, and **2-*t*-4**, had excellent antibacterial activity, which was comparable with that of vancomycin (VCM). The antibacterial effect of the three regioisomers was not significantly different. These findings indicate that it is not necessary to separate the regioisomers to study their biological activities. C_{60}-bis(2-alkyl-*N,N*-dimethylpyrrolidinium iodide) (**3**, **4**) and C_{60}-bis(*N*-alkyl-*N*-methylpyrrolidinium iodide) (**6**, **7**) also showed antibacterial activity, and **6** was the most effective. Moreover, these derivatives inhibited the growth of vancomycin-resistant *E. faecalis*.

In contrast to **2**, **3**, **4**, **6**, and **7**, derivatives with a long alkyl chain, **5** and **8**, had no antibacterial activity. In the respiratory chain inhibition, **5** and **8** had from none to slight activity. These results indicated that the mechanism of antimicrobial activity is a respiratory chain inhibition and that appropriate lipophilicity of the derivatives was suitable for the inhibition of the respiratory chain and for antibacterial activity.

Tsao et al. [9] reported the antibacterial activity of carboxy fullerenes. They have shown that its action is achieved by insertion into the cell walls and disruption of the cell structure. This mechanism is different from the respiratory chain inhibition. The bacteriostatic effect of fullerene derivatives has been reported by Bosi et al. [10], but the mechanism was not investigated.

Journal of Physics: Conference Series **159** (2009) 012003 doi:10.1088/1742-6596/159/1/012003

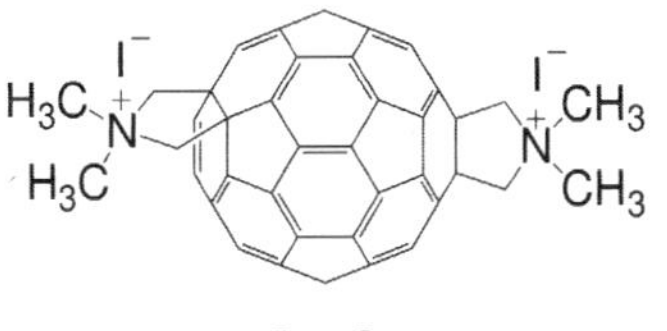

2-*t*-2

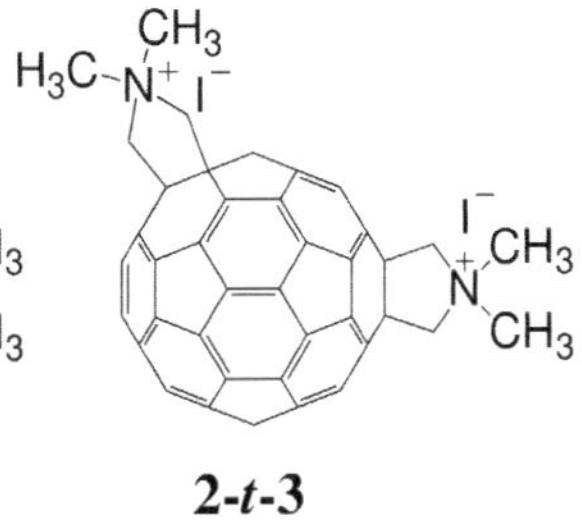

2-*t*-3

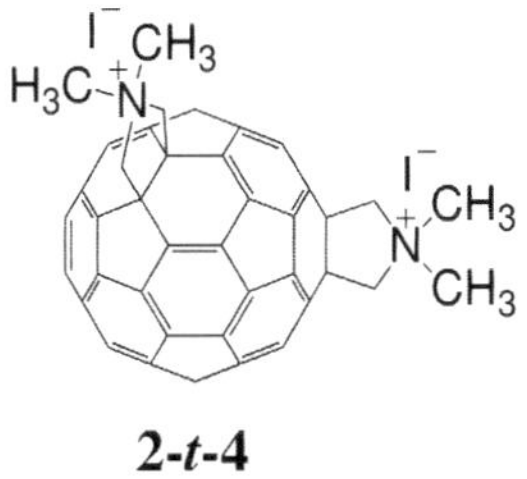

2-*t*-4

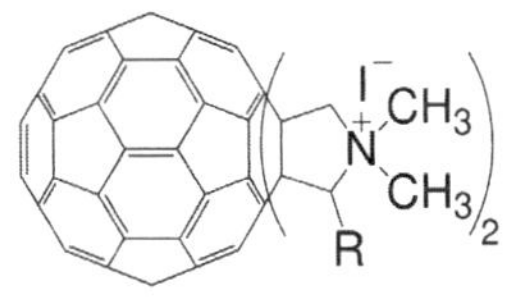

3: R = *n*-C_4H_9
4: R = *n*-C_6H_{13}
5: R = *n*-C_9H_{19}

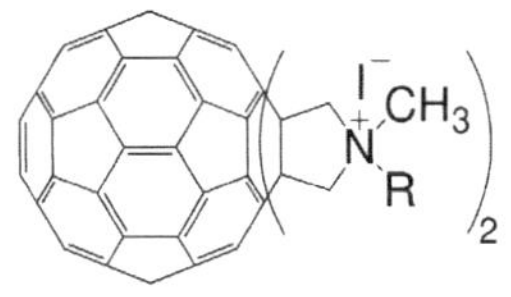

6: R = *n*-C_4H_9
7: R = *n*-C_7H_{15}
8: R = *n*-$C_{10}H_{21}$

Figure 7. Structure of C_{60}-bis(*N*,*N*-dimethylpyrrolidinium iodide) regioisomers (**2-*t*-2**, **2-*t*-3**, **2-*t*-4**) and alkylated C_{60}-bis(*N*,*N*-dimethylpyrrolidinium iodide) derivatives (**3** – **6**).

Table 1. Antibacterial activity of fullerene derivatives.

	MIC (μg/mL)									
	2-*t*-2	**2-*t*-3**	**2-*t*-4**	**3**	**4**	**5**	**6**	**7**	**8**	**VCM**
S.aureus 209P JC-1	1.56	0.78	3.12	6.25	3.12	>100	0.36	6.25	>100	1.56
S.aureus M133 (MRSA)	0.78	1.56	3.12	6.25	6.25	>100	0.36	6.25	>100	1.56
S.aureus M126 (MRSA)	3.12	1.56	3.12	6.25	3.12	>100	0.78	6.25	>100	1.56
S.eidermidis ATCC 14990	3.12	1.56	3.12	6.25	3.12	>100	0.78	6.25	>100	1.56
E.hirae ATCC 8043	12.5	6.25	6.25	6.25	25.0	>100	1.56	6.25	>100	3.12
E.faecalis W-73	12.5	6.25	6.25	6.25	12.5	>100	1.56	6.25	>100	3.12
E.faecium vanA (VRE)	12.5	6.25	6.25	12.5	6.25	>100	1.56	6.25	>100	>100
E.faecalis NCTC 12201 (VRE)	12.5	3.12	6.25	6.25	6.25	>100	1.56	6.25	>100	>100

4. Antiproliferative activity of cationic fullerene derivatives

The antiproliferative activities of fullerene derivatives were evaluated and compared with well-known anticancer agents by using a panel of 36 human cancer cell lines. The antiproliferative activity was measured by the MTT assay, giving the fullerene derivative concentration required for 50 % growth inhibition (GI_{50}). The average of the logarithm of GI_{50} (MID-GI_{50}) is presented in Table 2.

Fullerene derivatives **2** – **5** were effective in the antiproliferative activity, which was comparable to that of cisplatin (MID-GI_{50} = –5.30). In case of antibacterial activity, **5** was ineffective. It seems that the mechanism of antiproliferative activity is not the same as that of antibacterial activity and/or that the distribution of these derivatives in a mammalian cell is different from that in bacteria.

Tabata et al. have reported the anticancer activity of a fullerene derivative with laser photo-irradiation [4]. A laser photo-irradiated fullerene derivative produces active oxygen to kill cancer cells. The action was not replicated in the present experiment because photo-irradiation was not applied.

Table 2. Antiproliferative activity of fullerene derivatives.

compound	MID-GI_{50}
2-*t*-2	–5.44
2-*t*-3	–5.18
2-*t*-4	–5.18
3	–5.42
4	–5.48
5	–5.61
cisplatin	–5.30

5. Human immunodeficiency virus-reverse transcriptase inhibition activities of proline-type fullerene derivatives

HIV infection is one of the major causes of morbidity and mortality in the world. There are many anti-HIV agents, but their efficiency is not very high. Moreover, the emergence of drug-resistant mutant forms of HIV requires the development of effective and well-tolerated new remedies. These threats provide motivation to search for new types of lead compounds to be used as medicine against HIV infection. We therefore examined the HIV-RT inhibition activities of anionic (**1**), cationic (**2** – **8**), and proline-type (Fig. 8, **9** – **12**) fullerene derivatives.

The HIV-RT inhibition activity of the fullerene derivatives is listed in Table 3. All examined fullerene derivatives were more effective than the non-nucleoside analog of the HIV-RT inhibitor, nevirapine. Nevirapine, under the brand name Viramune®, is now used for HIV infection. Especially, the proline-type fullerene derivatives, **10** and **12**, strongly inhibited HIV-RT.

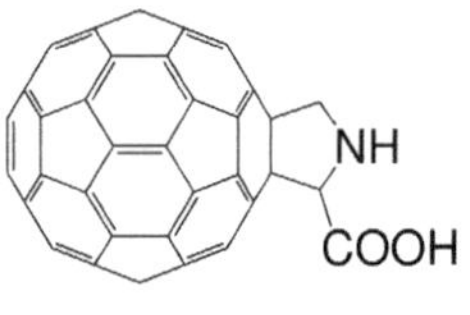

9

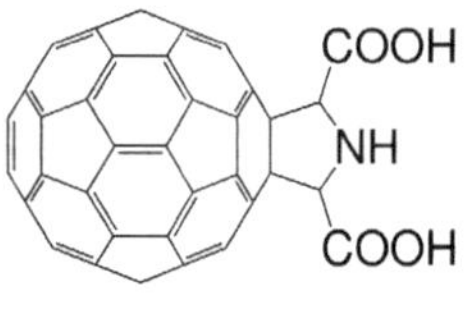

10

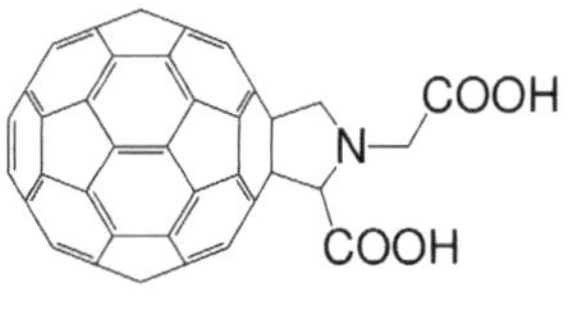

11

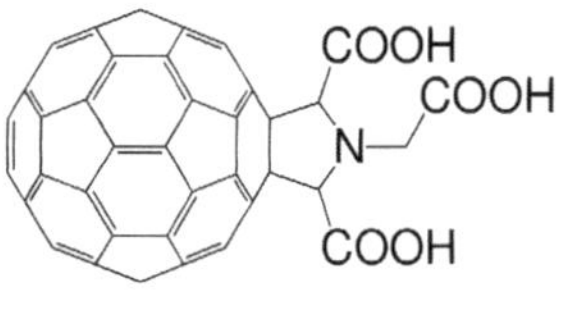

12

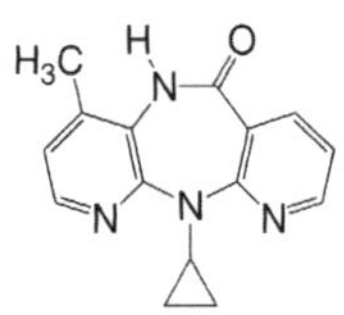

nevirapine

Figure 8. Structure of proline-type fullerene derivatives and nevirapine

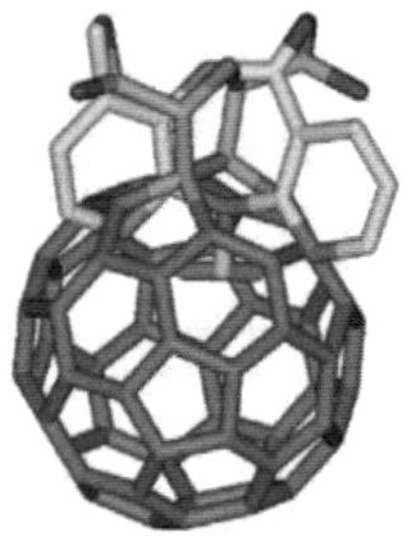

Figure 9. Superimposition of nevirapine and **10** at nevirapine-binding site of HIV-RT.

Docking simulation was performed by DS Modeling 1.2-SBD.

The X-ray crystal structure of the complex of HIV-RT and nevirapine has already been available (PDB ID; 3HVT). A docking simulation study of HIV-RT and fullerene derivative **10** was investigated, and **10** was able to fit the nevirapine-binding site of HIV-RT. Figure 9 shows a superimposition of the binding states of nevirapine and **10**. The pyrrolidine ring of **10** is located on the B-ring of nevirapine, and the position of C_{60} core is on the side of the cyclopropane ring of nevirapine.

Table 3. HIV-RT inhibition activity of fullerene derivatives.

compound	IC_{50} (μM)	compound	IC_{50} (μM)
1	1.2	**6**	2.5
2-*t*-2	1.8	**7**	1.9
2-*t*-3	1.7	**8**	8.9
2-*t*-4	1.9	**9**	0.15
3	1.1	**10**	0.022
4	0.84	**11**	0.073
5	0.59	**12**	0.029
		nevirapine	3.0

The HIV-RT inhibition activities were examined according to Mizrahi et al. [11]. One μL of a DMSO sample solution and 1 μL of HIV-RT (0.01 U/mL) were added to a 18 μL reaction mixture containing 50 mM Tris-HCl (pH 8.3), 30 mM NaCl, 10 mM $MgCl_2$, 5 mM dithiothreitol, 0.125 mg/mL poly(rA)•oligo$(dT)_{12\text{-}18}$, and 2.5 μM dTTP including ^{32}P-dTTP. It was incubated for 1 h at 37 °C. Then, 10 μL of the reaction mixture was taken and placed on a Whatman® DE81 chromatography paper. After the chromatography paper was dried, it was washed three times with a 0.5 M NaH_2PO_4 buffer (pH 7.0), 70 % ethanol, and ethanol. The radioactivity of the dried chromatography paper was counted with a liquid scintillator, and the HIV-RT activity was measured. IC_{50}s are the means of three experiments.

There are two major targets for anti-HIV agents, i.e., HIV-protease and HIV-reverse transcriptase. Molecular modeling studies have revealed that the C_{60}-core could fit into the large and highly hydrophobic substrate-binding site of HIV-protease. Indeed, some fullerene derivatives inhibited HIV-protease [12]. However, we propose that the fullerene derivatives also have HIV-RT inhibition activity. Bosi et al. have reported the anti-HIV activity of fullerene derivatives [13]. They speculated that the mechanism of anti-HIV activity is HIV-protease inhibition but lacked experimental evidence. Our HIV-RT inhibition results demonstrate another possible mechanism of the anti-HIV activity of fullerene derivatives.

6. Conclusion

We have reported the antioxidant, antibacterial, and antiproliferative activities of fullerene derivatives. With regard to the antioxidant activity, anionic fullerene derivatives such as **1** decreased the active oxygen toxicity by $O_2^{\bullet-}$ quenching. Anionic fullerene derivative **1** also relieved *E. coli* from the growth inhibition by $O_2^{\bullet-}$ generator. For the antibacterial and antiproliferative agents, cationic derivatives such as **2** to **8** were promising agents. Cationic fullerene derivatives were suggested to produce reactive oxygen species, and the respiratory chain was inhibited. The proline-type fullerene derivative, **11**, was an excellent lead compound for the anti-HIV agent. Fullerene derivatives have many interesting biological activities, as we and other groups have already reported. These activities depend on the properties of the fullerene core, while the substituents on the fullerene core control and modify the biological activities of fullerene derivatives.

Acknowledgements
The authors wish to thank Dr. T. Yamori for the measurement of antiproliferative activity of fullerene derivatives. This work was supported in part by a grant from the Science Research Promotion Fund of the Japan Private School Promotion Foundation.

References
[1] Oberdörster G 2004 *Environ. Health Prospect.* **112** 1058
[2] Usenko C Y, Harper S L and Tanguay R L 2008 *Toxicol. Appl. Pharmacol.* **229** 44
[3] Deguchi S, Mukai S, Tsudome M and Horikoshi K 2006 *Adv. Mater.* **18** 729
[4] Tabata Y, Murakami Y and Ikeda Y 1997 *Jpn. J. Cancer Res.* **88** 1108
[5] Kato H, Kanazawa Y, Okumura M, Taninaka A, Yokawa T and Shinohara H 2003 *J. Am. Chem. Soc.* **125** 4391
[6] Schinazi R F, Sijbesma R, Srdanov G and Hill C L and Wudl F 1993 *Antimicrob. Agents Chemother.* **37** 1707
[7] Chiang L Y, Lu F -J and Lin J -T 1995 *J. Chem. Soc., Chem. Commun.* 1283
[8] Mashino T, Usui N, Okuda K, Hirota T and Mochizuki M 2003 *Bioorg. Med. Chem.* **11** 1433
[9] Tsao N, Luh T-Y, Chou C-K, Chang T-Y, Wu J-J, Liu C-C and Lei H-Y 2002 *J. Antimicrob. Chemother.* **49** 641
[10] Bosi S, Da Ros T, Castellano S, Banfi E and Prato M 2000 *Bioorg. Med. Chem. Lett.* **10** 1043
[11] Mizrahi V, Lazarus G M, Miles L M, Meyers C A and Debouck C 1989 *Arch. Biochem. Biophys.* **273** 347
[12] Friedman S H, Ganapathi P S, Rubin Y and Kenyon G L 1998 *J. Med. Chem.* **41** 2424
[13] Bosi S, Da Ros T, Spalluto G, Balzarini J and Prato M 2003 *Bioorg. Med. Chem. Lett.* **13** 4437

Field-Effect-Transistor Characteristics of Solvate C_{60} Fullerene Nanowhiskers

Y Ochiai[1], K Ogawa[1], N Aoki[1] and J P Bird[2]

[1]Graduate School of Advanced Integration Science, Chiba University, Chiba, Japan
[2]Department of Electrical Engineering, University at Buffalo, SUNY, Buffalo, USA

E-mail: ochiai@faculty.chiba-u.jp

Abstract. We demonstrate the operation of C_{60} fullerene nanowhisker (FNW) field-effect transistors (FETs), realized by dissolving C_{60} powder in either *m*-xylene or chlorobenzene solvents and performing liquid-liquid interface precipitation. There exists a clear dependence on solvent species for the FET performance. In contrast to prior studies, in which the FETs were investigated under vacuum conditions to avoid introducing crystalline disorder, we achieve comparable transistor operation for FETs under a N_2 atmosphere. The finding that solvate FNWs may be used to implement FETs could have important technological implications, by allowing practical applications of these materials to be developed.

1. Introduction

Organic thin-film field effect transistors (FETs) have been widely investigated in recent years, from the perspective of developing near-future device applications [1]. Among such materials, however, only a few show *n*-type characteristics in FET operation. C_{60} is one of these materials, and improvements in its FET characteristics are therefore desired [2]. Recently, a novel approach for synthesizing fibrous C_{60} crystals has been developed by using the method of liquid-liquid interface precipitation (LLIP). This yields C_{60} fullerene nanowhiskers (FNWs) with uniform nanoscale cross-section, over lengths that can easily exceed several microns [3]. Previously, we reported on the fabrication of FET structures incorporating either multiple [4] or single [5] C_{60} FNWs. For practical applications, it is desirable to be able to operate such FETs under ambient conditions. Unfortunately, however, the performance of C_{60}-based FETs is known to degrade on exposure to air, since adsorbed oxygen creates a deep trapping level that suppresses carrier transport [6]. Consequently, to date we have only succeeded in measuring the electrical characteristics of FNW FETs under vacuum conditions, after performing thermal annealing to remove oxygen and other adsorbates. One of the characteristics of C_{60} FNWs synthesized by LLIP is that the as-grown, solvate C_{60} FNWs exhibit a hexagonal crystalline structure, in contrast to the face centered cubic (fcc) structure that is typical of C_{60} films at room temperature (RT). The hexagonal structure is converted to the normal fcc one, however, after evaporating the solvent from the crystal, and this change in crystal structure is known to lead to the introduction of a high density of dislocations [7]. As a result, the C_{60} FNW FETs that we have studied to date have typically suffered from pronounced crystalline disorder. In this work, however, we demonstrate an approach that allows us to achieve FET operation of FNW FETs in a nitrogen environment, without the need for vacuum isolation, thereby opening up the possibility of developing technological applications based on these devices.

2. Sample and experimental setup

The fabrication and structural characterization of the C_{60} FNWs we study was described in detail in a previous publication [4]. The process starts by preparing pure C_{60} powder by sublimation, in a vacuum chamber at a partial vacuum of order 10^{-6} Torr. The powder was then transferred into the N_2 environment of a glove box without being exposed to air and was dissolved to saturation in either *m*-xylene or chlorobenzene, allowing us to compare the influence of these solvents on the resulting electrical characteristics. C_{60} NWs were then synthesized by the LLIP method, utilizing liquid phase separation between the C_{60} solution and isopropyl alcohol, during which the mixture was stored in a refrigerator at ~ 300 K for one week. The FET samples were fabricated on a 600-nm thick SiO_2 layer on top of an n-type Si substrate and the source-drain electrodes were formed from Au with an adhesion layer of Ti. The device had a channel length (contact separation) of 5 μm and the FETs were formed by bridging these electrodes with several FNWs (Fig. 1 inset, for further information on this process, see [5]). To prevent contamination of the nanowhiskers by ambient oxygen, the critical steps of the device processing were prepared in a glove box filled with N_2 gas. Subsequently, the FETs were installed into the chamber of a probe station without exposure to air. This allowed us to measure the room-temperature electrical properties of the solvate nanowhiskers directly. We were also able, however, to reduce the solvent content of these NWs, by pumping the sample chamber down to a vacuum of ~10^{-6} Torr. Under these conditions, we measured the characteristics of the dried FNWs, both before and after thermal annealing for a day at ~440 K. In experiments where the sample chamber was pumped to vacuum, a change of the crystalline structure of the C_{60} FNWs was confirmed by X-ray powder diffraction. As reported previously [4,5], this change involves a transition from a well-defined hexagonal structure [8] to an fcc one whose diffraction peaks are broadened, indicating that the dried FNWs do not preserve the good crystallinity characteristic of the solvate condition. This change in crystalline character occurs even though the FNWs maintain their overall hexagonal cross-section [4, 5].

3. Results and discussion

In Fig. 1, we show the gate-voltage (V_g) characteristic of a C_{60} FNW FET, which was obtained by using the conducting substrate of the device as a back-gate (this accounts for the large values of this voltage). The LLIP process used to synthesize the FNW in this case involved the use of *m*-xylene solvent, and the measurements of Fig. 1 were obtained under the solvate condition due to the presence

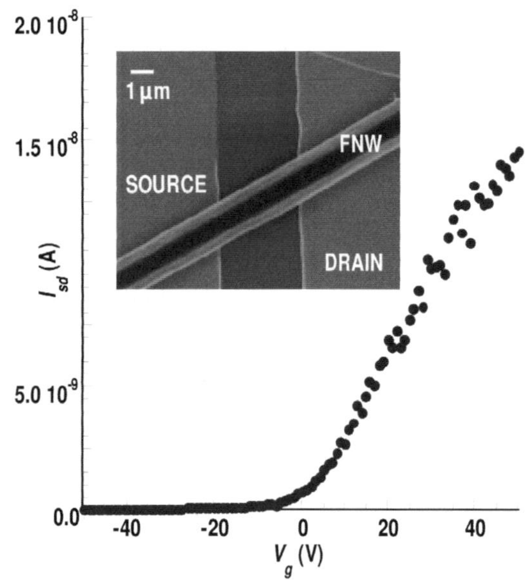

Figure 1. I_{sd}-V_g characteristic of a (*m*-xylene) C_{60} FNW FET measured under the solvate condition. The inset shows a scanning electron microscope image of the channel region of the FET.

of a N_2 atmosphere. From the V_g dependence of the source-drain current (I_{sd}), it can be inferred that the threshold voltage (V_t) and field-effect mobility (μ) are ~0 V and 4 × 10^{-3} cm^2/Vs, respectively. (We determine the field mobility from the following relation for the source-drain current: $I_{sd} = (\mu W C_o/L)(V_g - V_t - \frac{1}{2}V_{sd})V_{sd}$, where W, L, and C_o are the FET channel width, length, and capacitance, respectively [5]). This device shows a normal-on characteristic similar to our former reports, with a mobility only a little lower than that which we have previously found [4] for dried FNW FETs (see Table 1). This represents the first important result of our study, since it demonstrates the possibility of obtaining FET action from FNWs under other than vacuum conditions.

Subsequent to the measurements of Fig. 1, the probe-station chamber was next evacuated to dry out the FNWs. After measuring the resulting FET characteristics, the device was then annealed at 440 K for 24 hours and cooled back down to RT. The resulting gate-voltage characteristics for these conditions are shown in Fig. 2. Note that the current in this figure is plotted on a logarithmic scale, for the solvate (curve (a)), dried (curve (b)), and dried and annealed (curve (c)), conditions. From these data it can be determined that I_{sd} (V_g = 0) increases on evaporating the solvent, while the on/off ratio decreases from four- to two-orders of magnitude. On the other hand, we find that the mobility is improved to 10^{-2} cm^2/Vs as a result of the annealing, an improvement of roughly a factor of two over the solvate condition.

We have also performed the same procedures as those described above, in this case for an FET whose channel was formed from FNWs realized using chlorobenzene as the solvent. The corresponding variations of $I_{sd}(V_g)$ measured in this case are shown in Fig. 3. From these data, we are again able to infer that the solvate FNW yields transistor characteristics that compare favourably with those obtained under vacuum conditions. The on-off ratio is a factor of ten larger in the solvate condition, but the inferred mobility is actually higher. These results are also summarized in Table 1, based on inspection of which we can make some further comments. In the solvate condition, it appears the mobility of the chlorobenzene-derived FNW is significantly enhanced over that of the *m*-xylene-derived one. Although the reasons for this are not completely understood, an important factor may be the different polar nature of the two solvents we have used. Specifically, Cl should act as a weak acceptor in chlorobenzene, while CH_3 acts as a donor in *m*-xylene with benzene. These results therefore suggest the importance of utilizing an appropriate solvent as a means to enhance the electrical characteristics of the solvate FNWs. The other noteworthy feature of Table 1 concerns the characteristics achieved after drying and annealing the FNWs. Since in this case the solvent is

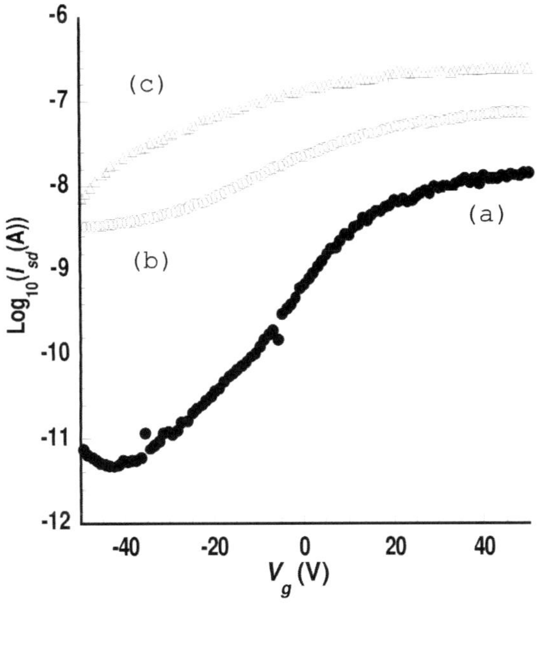

Figure 2. I_{sd}-V_g characteristic (with current plotted on a logarithmic scale) of the FNW FET of Fig. 1, measured at room temperature under three different conditions: N_2 atmosphere in solvate condition (curve (a), filled circles), dried in vacuum (curve (b), open circles), and dried in vacuum and annealed (curve (c), open triangles).

removed from the nanowhisker one might expect to obtain the same electrical characteristics for FNWs produced with either solvent. Table 1 confirms that the mobility and on-off ratios are the same in both cases. The threshold voltages are different, however, reflecting the unintentional doping induced by the different solvents.

Turning to the physical implications of our results, the observation of significant changes in the threshold voltage of both devices (Table 1) indicates that carriers are introduced into the FNWs, independent of the gate voltage, by evaporating the solvent. This may possibly be related to the introduction of a high density of dislocation and disorder [7], which could result in the creation of additional carriers in the FNWs. The low values of the mobility that we infer for both devices in the solvate condition appear reasonable, particularly in the case of *m*-xylene, since solvent molecules intercalated into the C_{60} crystal should extend the lattice constant, thereby decreasing the carrier hopping probability [9]. It is clear, on the other hand, that the mobility is not changed so systematically by removing the solvent from the nanowhiskers (compare the two different devices in Table 1). This may indicate some competition between the effect of eliminating the intercalated solvent molecules, and the ensuing increase in crystalline disorder when the crystalline structure of the FNWs undergoes a transition to the fcc form.

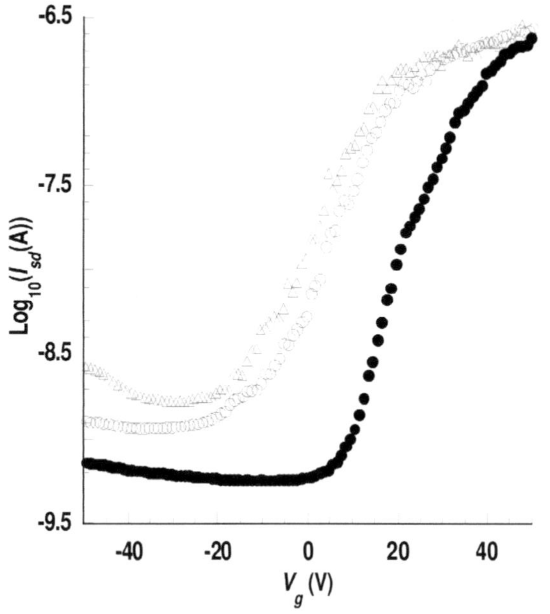

Figure 3. I_{sd}-V_g characteristic (with current plotted on a logarithmic scale) of a (chlorobenzene) C_{60} FNW FET, measured at room temperature under three different conditions: N_2 atmosphere (solvate condition, filled circles), dried in vacuum (open circles), and dried in vacuum and annealed (open triangles).

Table 1. Caption: Summary of the results of measurement for the *m*-xylene and chlorobenzene C_{60} FNW FETs under different atmospheric conditions. The on-off ratio is determined as the ratio of currents at $V_g = \pm 40$ V.

Solvent	Solvate Condition			Vacuum, Before Anneal			Vacuum, After Anneal		
	μ (cm^2/Vs)	V_t (V)	On/Off Ratio	μ (cm^2/Vs)	V_t (V)	On/Off Ratio	μ (cm^2/Vs)	V_t (V)	On/Off Ratio
***m*-xylene**	4×10^{-3}	0	$\sim 10^4$	$\sim 10^{-2}$	< -40	$\sim 10^2$	$\sim 10^{-2}$	-13	$\sim 10^2$
Chlorobenzene	5×10^{-2}	20	$\sim 10^3$	$\sim 10^{-2}$	0	$\sim 10^2$	$\sim 10^{-2}$	0	$\sim 10^2$

4. Conclusion

In conclusion, we have demonstrated the operation of C_{60} FNW FETs under a N_2 environment, and have been able to achieve reasonable transistor action without the need to operate these devices under stringent vacuum conditions. The crystalline structure of the FNWs changes from an oriented hexagonal form to a disordered fcc one on evaporating the solvent. Accompanying this, resulting dislocations in the crystalline structure are believed to introduce carriers into the FNWs, thereby modifying the threshold gate voltage. The finding that solvate FNWs may be used to implement FETs could have important implications by allowing for technological applications of these materials to be developed. Future work should therefore focus on exploring suitable solvents for FNW fabrication to yield optimal electrical characteristics.

Acknowledgments

The work was supported in part by Grants-in-Aid for Scientific Research from the Japan Society for the Promotion of Science (19054016, 16656007 and 16206001).

References

[1] Dimitrakopoulos C D and Marlenfant P L L 2002 *Adv. Mater.* **14** 99
[2] Haddon R C, Perel A S, Morris R C, Palstra T T M, Hebard A F and Fleming R M 1995 *Appl. Phys. Lett.* **67** 121
[3] Miyazawa K, Kuwasaki Y, Obayashi A and Kuwabara M 2002 *J. Mater. Res.* **17** 83
[4] Ogawa K, Kato T, Ikegami A, Tsuji H, Aoki N, Bird J P and Ochiai Y 2006 *Appl. Phys. Lett.* **88** 112109
[5] Ogawa K, Kato T, Ikegami A, Tsuji H, Aoki N, Bird J P and Ochiai Y 2006 *J. Phys. Conf. Series* **38** 1
[6] Horiuchi K, Nakada K, Uchino S, Hashii S, Hashimoto A, Aoki N, Ochiai Y and Shimizu M 2002 *Appl. Phys. Lett.* **81** 1911
[7] Miyazawa K, Minato J,Yoshii T, Fujino M and Suga T 2005 *J. Mater. Res.* **20** 688
[8] Ramm M, Luger P, Zobel D, Duczek W and Boeyens J C A 1996 *Crystal. Res. Technol.* **31** 43
[9] Horiuchi K, Uchino S, Hashii S, Hashimoto A, Kato T, Sasaki T, Aoki N and Ochiai Y 2004 *Appl. Phys. Lett.* **85** 1987

Hierarchically Organized Soft-Materials Based on Fullerenes

Takashi Nakanishi*[,1,2,3]
[1] Organic Nanomaterials Center, National Institute for Materials Science (NIMS)
1-2-1 Sengen, Tsukuba 305-0047, Japan.
[2] MPI-NIMS International Joint Laboratory, Department of Interfaces, Max Planck Institute of Colloids and Interfaces
Research Campus Golm, Potsdam 14424, Germany.
[3] PRESTO, Japan Science and Technology (JST)
4-1-8 Honcho Kawaguchi, Saitama 332-0012, Japan.

E-mail: NAKANISHI.Takashi@nims.go.jp

Abstract. Simple chemical modifications of fullerene (C_{60}) with long aliphatic chains provide novel type amphiphilic molecules playing in organic solvents due to the two different intermolecular interactions, namely π-π on C_{60} and van der Waals interactions on aliphatic chain moieties, respectively, and open a door developing supramolecular soft-materials having hierarchically organized architectures, various morphologies and functions based on fullerenes. By tuning the length and number of aliphatic chains on the derivatives as well as experimental conditions such as solvents, temperature, substrates for preparation of the assemblies, the assembled fullerenes showed various faces such as creating of many unique-shaped objects with controlled their dimensionality. For instance, nanowires and thin disks with single bilayer thickness in nanometer size, globular, fibrous, conical objects in mesoscopic (sub-micrometer) scale and flower-shaped and direction-controlled spiral objects in micrometer scale are obtained. As bulk states, thermotropic liquid crystals and room temperature (isotropic) liquid fullerenes are interestingly prepared from this molecular designs and showed not only their fluid natures and comparably high carrier mobility as fullerene-based organic-semiconductor phenomena. In addition, nano-carbon superhydrophobic surface with fractal morphology of the two-tier roughness on a nano- and microscopic scale was created from one of the supramolecular objects. The all of hierarchical supramolecular assemblies describing in this review is derived from fine-tuning intermolecular interactions of fullerene derivatives bearing long aliphatic chains.

1. Introduction

The motivation in this study is how to bridge "supramolecular chemistry" and "material sciences". The former is more related to organic chemistry, colloids and interfaces chemistry based on molecular organizations, while the latter one is required intrinsic functionalities, processability and stability of materials. In fact, nowadays there are many researches towards to development of field emission transistors (FET), organic light emitting diodes (OLED) and so forth containing organic substances [1, 2]. In most cases, conductive or semiconductive organic molecules or polymers are applied in these purposes. However, there are rarely considered that prepared (designed and synthesized) molecules can be hierarchically assembled into desired morphologies using a strategy of the supramolecular

* Dr. Takashi Nakanishi. (NIMS, MPI, and JST)

chemistry and became multi-tasked materials with opto-/electric-properties [3]. The selected candidate for the requirements in this study is fullerene (C_{60}). The C_{60} and the derivatives are one kind of promising organic materials because of the electron-transport property (*n*-type organic semiconductor), rich electrochemistry, photo-responsibility (light synthetic model materials), bio-activity and so on [4, 5]. Although there have been many researches studied on supramolecular chemistry of C_{60}, for instance, formation of the self-organized structures, liquid crystalline materials and molecular recognitions, there are only few reports concerning the properties and applications of supramolecular assemblies of fullerenes [6, 7]. With a view to regulation of their assembled superstructures, conventional amphiphilic approaches, which are based on the substitution of hydrophilic moieties at the intrinsically hydrophobic C_{60}, has been applied in many cases [8-11]. However, controlling the shape of the assemblies is not well-established so far, and only spherical vesicles or tubular assemblies are obtained in this strategy.

Herein, in this review, the author's recent studies related development of novel type soft-materials based on supramolecular chemistry using unique fullerene derivatives bearing long aliphatic chains as the molecular constituent are described. Fulleropyrrolidines functionalized with a multi(alkyloxy)phenyl group (Figure 1) satisfy some of requirements to be better soft-materials such as a high C_{60} content in the molecule, a predictable compact packing, a highly ordered mesophase and controllable the dimensionality of the self-organized structures. Moreover, fulleropyrrolidines form stable species under electrochemical reduction of the C_{60} cage and thus as redox-active soft-materials. The developed supramolecular fullerene nano- and micro-architectures are utilizing the intermolecular forces introduced by C_{60} (π-π) and aliphatic chain interaction (van der Waals). A delicate balance between the π–π and van der Waals interactions in the assemblies leads to a wide variety of supramolecular architectures.

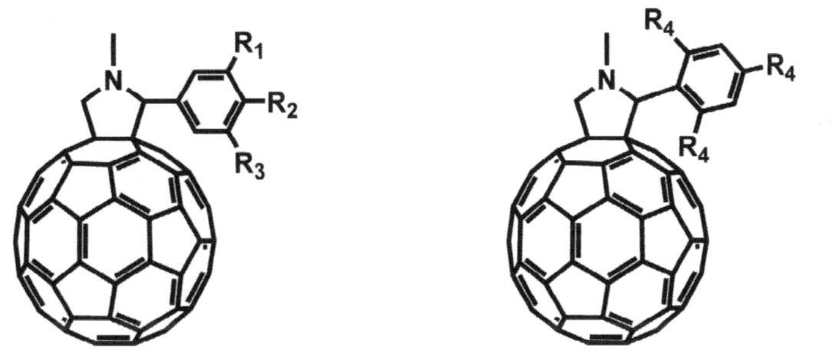

1 : $R_1 = R_2 = R_3 = OC_{20}H_{41}$
2 : $R_1 = R_2 = OC_{20}H_{41}$, $R_3 = H$
3 : $R_1 = R_3 = H$, $R_2 = OC_{20}H_{41}$
4 : $R_1 = R_2 = R_3 = OC_{16}H_{33}$
5 : $R_1 = R_2 = OC_{16}H_{33}$, $R_3 = H$
6 : $R_1 = R_3 = H$, $R_2 = OC_{16}H_{33}$
7 : $R_1 = R_2 = R_3 = OC_{12}H_{25}$

8 : $R_1 = R_2 = R_3 = OC_{20}H_{41}$
9 : $R_1 = R_2 = R_3 = OC_{16}H_{33}$
10 : $R_1 = R_2 = R_3 = OC_{12}H_{25}$

Figure 1. Chemical structures of fullerene derivatives employing in this study.

2. Dimension Controllable Supramolecular Fullerene Assemblies

Fullerene derivatives bearing three alkyloxy aliphatic chains (Figure 1, **1-7**) were prepared by refluxing the corresponding alkyloxybenzaldehyde with *N*-methylglycine and C_{60} in dry chlorobenzene, and unambiguously characterized by ^{1}H and ^{13}C NMR, FT-IR and UV-visible spectroscopy, as well as MALDI-TOF mass spectroscopy [12-16]. Fullerene derivative (**4**), a fulleropyrrolidines functionalized with a 3,4,5-(hexadecyloxy)phenyl group showed polymorphism [12, 17, 18], different well-defined architectures, on the supramolecular assemblies in various solvent systems (Figure 2). The sample preparation is quite simple as follows. Self-assembled supramolecular objects were prepared by evaporation to dryness of a 1 mL chloroform solution of **4** ([**4**] = 1.0 mM) followed by addition of 1 mL of the respective solvents. Subsequent heating 60-70 °C for 2 hours resulted in light brown mixtures, which were aged at 20 °C for 24 hours, prior to microscopy examination. Solvents of variable polarity used in these experiments included 2-propanol/toluene mixtures, 1-propanol, H_2O/THF mixture, 1-butanol, 1,4-dioxane and enantiopure 2- (*S*)- or (*R*)-butanol. Mixtures in a 2-propanol/toluene solution were subsequently cooled to 5 °C for at least 12 hours and ultrasonication was performed at 5 °C for 1 hour. 1,4-Dioxane solution of **4** was also subsequently cooled to 5 °C for 12 hours, resulting in formation of dark brown colored precipitates. As shown field emission scanning electron microscopic (SEM) images in Figure 2, fullerene derivative **4** self-assembles into hierarchically ordered nano- and micro-architectures, such as vesicles with an average diameter of 250 nm in 1:1 2-propanol/toluene mixture at room temperature (Figure 2a), fibers with partially twisted tapes in 1-propanol (Figure 2b), conical objects with diameter of 60 nm perforated at the cone apex in 1:1 H_2O/THF mixture (Figure 2c), microspheres with diameters of 1-2 μm (Figure 2d) and maracas-like (Figure 2e) under ultrasonication at 5 °C for 60 min and 30 min, respectively. Windmill-like sheets architectures were obtained in a 1:2 2-propanol/toluene mixture at 20 °C as brown precipitates (Figure 2f). Uncommon self-organized architectures of **4**, jellyfish-like, are seen from objects prepared in 1-butanol at 20 °C (Figure 2g). Furthermore the left-handed spiral microstructures were obtained from 2-(*R*)-butanol (Figure 2h), whereas 2-(*S*)-butanol provided right-handed spiral objects (Figure 2i) of 3-6 μm in diameters.

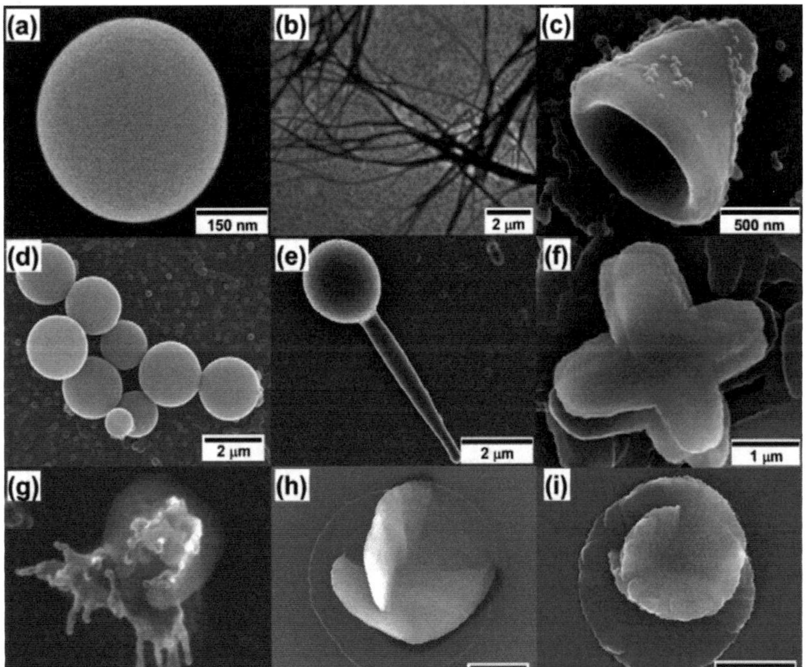

Figure 2. SEM images of spherical (a), fibrous (b), conical (c), microspherical (d), maracas-like (e), windmill-like (f), jellyfish-like (g), left-handed (h) and right-handed (i) spiral objects of **4** obtained various solvent conditions. Detail conditions are described in the main text. Reprinted with permission from [12], Nakanishi T et al, 2005 *Chem. Commun.* 5982. © 2005, Royal Society of Chemistry. [17], Nakanishi T et al, 2007 *Small* **3**, 2019. © 2007, WILEY-VCH.

Even flat disk-shape objects are one of the attractive morphologies in surfactant assemblies [19], normally, two-dimensional planar sheet assemblies turn into tubular assemblies with cigar-like scrolls via rolling up of the sheet in order to minimize the total elastic energy [20]. In this current study, however, planar windmill-like sheet assemblies overlie each other without scrolling the sheets and also the spiral objects are originating from disk-shaped precursor assemblies. Further studies for understanding the formation mechanism, including theory, of the 2-dimensional assemblies and the reasons of the tetra-clefts formation, the anisotropic growth as well as of the influence of the chiral solvent at the molecular level on the microscopic spiral morphogenesis are still required. Slight differences of the aliphatic chain length and numbers of the derivatives cause great morphology changes in their supramolecular assemblies. Remarkably, these reference derivatives **5-7** [14, 18] never showed polymorphs in various organic solvents as seen in **4** [12, 17, 18].

Interestingly C_{60} itself also forms various shaped assemblies under different sample preparation procedures such as liquid-liquid interfacial precipitation [21] and solvent-induced reprecipitation process [22]. These researches proved three dimensional π-π interactions on the C_{60} cores offer them assembling into different single crystalline arrangements and resulting in various shaped C_{60} microcrystalline structures. Research examples described here open a new path how to prepare supramolecular soft-materials based on fullerenes together with desired morphologies.

3. Supramolecular Shape-Shifter from Nano-Disk to Micro-Flower Architectures

To further diversify the library of supramolecular fullerene objects, for instance, with fractal shape, the understanding of the formation mechanism of self-assembled microscopic objects was become more interested in. In 1,4-dioxane solution of **4** at 20 °C, disks with diameters of 0.2–1.5 μm (Figure 3a) and a 4.4 nm thickness (Figure 3b), confirmed by atomic force microscopy, were obtained [12]. The thickness is good agreement with the thickness of aliphatic chain-interdigitated single bilayer (see schematic model in Figure 3c). In fact, structural analyses of the assembled objects or cast thin films of **4** by cryo-TEM, DSC, FT-IR, UV-Vis spectroscopies and XRD indicated multi-bilayer organized structures with the layer distance around 4.4 nm are contributing into the supramolecular assemblies of **4**. In other words, the hierarchical organization of **4** consists of interdigitated bilayer-based nanostructures, which further assemble into microscopic architectures such as flowerlike objects [17] (Figure 3f). The flowerlike assemblies obtained as precipitates after cooling down to 5 °C the 1,4-dioxane solution from at 20 °C and several micrometers in size (3-10 μm) with crumpled sheet- or flakelike nanostructures several tens of nanometers in thickness.

Direct observation of the intermediate structure forming from nano objects to microscopic objects helps to deep understanding of the formation mechanism [17]. Rapid cooling of the homogeneous 1,4-dioxane solution of **4** from 60 °C to 5 °C yields intermediate assembled structures of the flowerlike objects. The pre-formed disk objects seen at 20 °C loosely roll up at the edges (Figure 3d). The rolling distortions take place in every quarter of the disk, resulting in square-shaped objects having four corners with conical shapes developed by the encounters of the rolled or folded edges on the disks. Such bending of a thin sheet is, in general, energetically more favourable than stretching it. When the rolling-up proceeds continuously, spatial congestions occur at the four corners, resulting in crumpling, bending, stretching, and fracture of the disks (Figure 3e). After these transformations are complete, the bilayer growth at the edges continues, fixing the spatial conformation of the crumpled sheets, and this transformation leads to the final form of the flower-shaped objects (Figure 3f). These results indicate the transformation (shape-shift) mechanism from molecular assembled bilayer disks into microscopic flower-shaped objects based of fullerene derivative **4**. Importantly, the ability to self-organize well-defined, discrete microscopic objects with complex morphologies may stimulate further advances in component design and self-assembly theory that go beyond simple morphologies, such as micelles, vesicles and tubes.

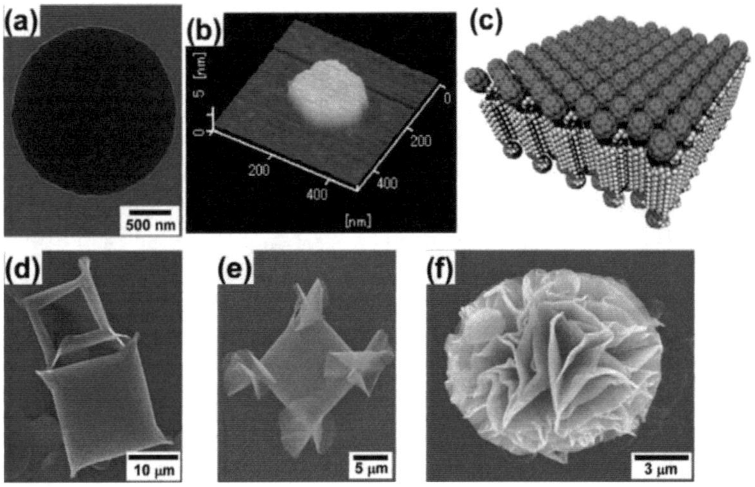

Figure 3. SEM (a) and AFM (b) images of disk-shaped assemblies formed in of **4** as a precursor for the flower formation. The proposed structural model of bilayer assembly (c). SEM images of square-shaped objects loosely rolled up in every corner (d), the further rolled-up objects of crumpled structures at the four corners (e), and the final flower-shaped objects (f). Reprinted with permission from [17], Nakanishi T et al, 2007 *Small* **3**, 2019. © 2007, WILEY-VCH.

4. Superhydrophobic Objects Based on Supramolecular Assembles of Fullerenes

Molecular self-organization of a fullerene derivative (**1**), a fulleropyrrolidines functionalized with a 3,4,5-(eicosyloxy)phenyl group, leaded to macroscopic globular objects with a two-tier roughness on the micro- and nanoscopic length scale (Figure 4a) precipitated from 1,4-dioxane solution after heating and subsequently cooling to room temperature [23]. Surface resulting from simple casting of these objects are superhydrophobic with a water contact angle of 152° (Figure 4b) and surprisingly durable towards polar organic solvents, and acidic/basic aqueous media, as well as heating [23]. The durability of the fabricated structures is related to the apolar nature of the constituents in combination with strong π-π and van der Waals interactions of C_{60} moieties and aliphatic chains.

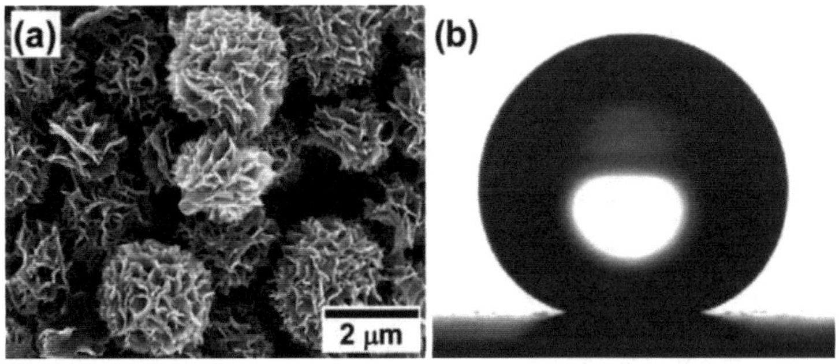

Figure 4. SEM image (a) of globular objects of **1** precipitated from 1,4-dioxane. A photograph of water droplet (b) on the surface (contact angle of 152°). Reprinted with permission from [23], Nakanishi T et al, 2008 *Adv. Mater.* **20**, 443. © 2008, WILEY-VCH.

It is also the fractal morphology and two-tier roughness of the globular objects of **1** on both the nano- and microscopic length scale, originating from the hierarchical self-organization starting from bilayer formations, which adds to the superhydrophobicity of the resulting surfaces. In contrast to most other methods used for the preparation of superhydrophobic surfaces [24], the prepared thin films described here are readily re-used or recovered by simply dissolving them in chloroform. This is an advantage for superhydrophobic surfaces created from hierarchical supramolecular assemblies.

5. Thermotropic Liquid Crystals of Fullerenes

Fullerene C_{60} is one kind of promising electron-transport materials as an *n*-type organic semiconductor [25]. Although various morphologies of C_{60}-containing liquid crystals (LC) have been studies in the past decade [26], there is only an example the carrier mobility of fullerenes in the LC state has been evaluated very recently [27]. However, the conducted value, ~10^{-4} cm^2/Vs, was not large enough for charge transporting materials, probably due to the moderate ordered structure and the low C_{60} content in the mesophase. Our fullerene derivatives satisfy the requirements for a high carrier mobility in the C_{60}-containing mesomorphic materials: a high C_{60} content up to 50 % (in the case of **2**) and a highly ordered mesophase. Moreover, fulleropyrrolidines form stable species under electrochemical reduction of the C_{60} cage and thus act as redox-active mesomorphic materials [28].

Thermal properties of **1** have been examined by a combination of DSC, POM and XRD. Thermotropic mesophase of **1** is seen in temperature range between 62 to 193 °C, and shows optical texture under polarized optical microscope (Figure 5a), which exhibit birefringence and confirmed the fluid nature. The XRD pattern for **1** at 185 °C shows a strong peak at $2\theta = 1.58°$ assigned (0 0 1), *d* spacing = 5.59 nm, accompanied by higher ordered peaks up to (0 0 14) (Figure 5b). In addition, a broad halo centred at $2\theta \approx 19°$ is assigned to the molten alkyl chains. These higher-degree peaks in the XRD pattern reveal a long-range ordered lamellar mesophase comparable to ordered smectic phase. Derivatives **2** and **4** also showed similar LC characteristics having long-range ordered lamellar mesophase. In other cases of derivatives **3**, **5**, **6**, **7** were found to be non-mesomorphic. The mesomorphic fullerenes feature reversible electrochemistry ($E_{red,1} = -0.70$ and $E_{red,2} = -0.87$ V) and a comparably high electron carrier mobility, ~3×10^{-3} cm^2/Vs for **1** at 120 °C evaluated by a conventional time-of-flights setup (ToF), making them attractive components for fullerene-based soft materials [28].

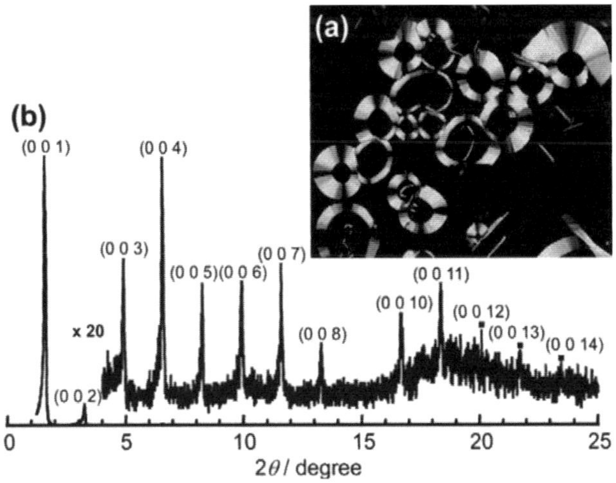

Figure 5. POM texture of mesophase of **1** at 190 °C (a) and XRD patterns of **1** at 185 °C (b). Reprinted with permission from [28], Nakanishi T et al, 2008 *J. Am. Chem. Soc.* **130**, 9236. © 2008, American Chemical Society.

6. Room Temperature Liquid Fullerenes

In the course of above studies, it was discovered serendipitously that fulleropyrrolidines substituted with a 2,4,6-tris(alkyloxy)phenyl group (**8-10**) exhibit a fluid phase at room temperature (Figure 6a) [16]. The key point is to select, as the substituent group, a structure which is molecularly-designed in such a way that the alkyl chains spread independently, and cohesion of the fullerene moiety is skillfully suppressed. For derivatives **8-10** studied and over the measured frequency range of the rheological behavior, the loss modules *G*" is higher than the storage modules *G*' and suggests liquid-like behavior. As a result of a systematic investigation of the fluid behavior of liquid fullerenes and their functions, the viscosity of the liquid can be controlled by changing the length of the introduced alkyl chain has been discovered (Figure 6b). Interestingly, the longer alkyl chain length is reducing efficiently the viscosity, which is completely opposite trend of a phenomenon of viscosity in normal alkane molecules.

Moreover, because the liquid fullerenes retain the characteristic properties of fullerenes, the liquid is electrochemically active. Cast films of the liquid fullerenes showed two-reductive events of C_{60}. As a further advantage of the liquid, this material also has relatively high hole mobility estimated from transient photocurrent measurements, ~3×10^{-2} cm^2/Vs for **8** at 20 °C [16]. These features make it an extremely attractive novel carbon material for future applications.

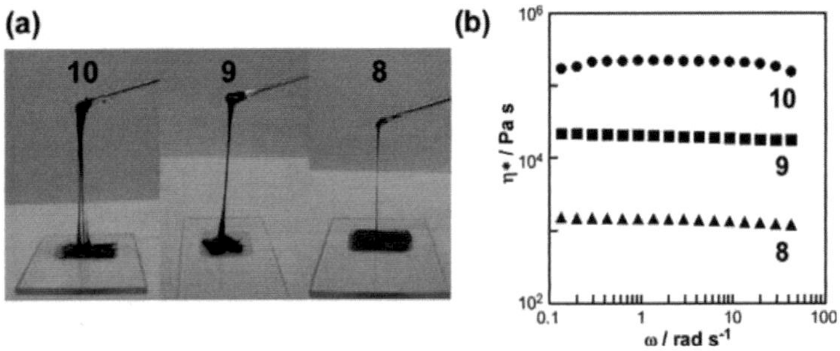

Figure 6. Photographs (a) of room temperature liquid fullerenes (**8-10**) and (b) results of rheology (complex viscosity) experiments of **8-10** at a strain amplitude of 0.01. Reprinted with permission from [28], Nakanishi T et al, 2006 *J. Am. Chem. Soc.* **128**, 10384. © 2006, American Chemical Society.

7. Conclusion

By simple molecular design, combination of C_{60} (π-π) and aliphatic chains (van der Waals), dimensionally regulated supramolecular objects were successfully prepared and enabled us to construct various organic soft-materials (superhydrophobic surfaces, thermotropic liquid crystals, room temperature liquid fullerenes, etc.) based on fullerenes with unique morphologies and the electric functions. Additional parameter to control the self-organization of molecules, such as interaction between molecules and substrate, can be also applied in the present fullerene derivatives, resulting in perfectly straight fullerene nanowires on graphite [13, 15]. The systems described here open the door for novel methodologies developing supramolecular fullerene soft-materials with controlled dimensionality and the desired fullerene functionalities.

Acknowledgments

This work was supported, in part, by a Grant-in-Aid for "Young Scientists (B) (No. 17750140 and 19750124)" from the Ministry of Education, Sciences, Sport and Culture, Japan, "PRESTO", JST, Japan, Iketani Science and Technology Foundations (171041A and 191127A), and Ishikawa Carbon Foundation. T. N. is grateful to Prof. H. Möhwald, Dr. D. Kurth, Dr. Y. Shen, Dr. J. Wang, Dr. N.

Miyashita MPI Colloids and Interfaces, Dr. K. Ariga, Dr. T. Michinobu, Dr. J. P. Hill, Dr. N. Shirahata, NIMS, Dr. H. Takahashi, Prof. T. Teranishi, Univ. of Tsukuba, Dr. K. Yoshida, Kyoto Univ. and Dr. M. Funahashi, Tokyo Univ. for helpful discussion and some of experimental contributions.

References

[1] Martín N, Sánchez L, Herranz M Á, Illeacas B and Guldi D M 2007 *Acc. Chem. Res.* **40** 1015
[2] Hutchison K, Gao J, Schick G, Rubin Y and Wudl F 1999 *J. Am. Chem. Soc.* **121**, 5611
[3] Yamamoto Y, Fukushima T, Suna Y, Ishii N, Saeki A, Seki S, Tagawa S, Taniguchi M, Kawai T and Aida T 2006 *Science* **314** 1761
[4] Martín N 2006 *Chem. Commun.* 2093
[5] Nakamura E and Isobe H 2003 *Acc. Chem. Res.* **36** 807
[6] Nierengarten J-F 2000 *Chem. Eur. J.* **6** 3667
[7] Imahori H, Norieda H, Yamada H, Nishimura Y, Yamazaki I, Sakata Y, Fukuzumi S 2001 *J. Am. Chem. Soc.* **123** 100
[8] Cassell A M, Asplund C L and Tour J M 1999 *Angew. Chem. Int. Ed.* **38** 2403
[9] Zhou S, Burger C, Chu B, Sawamura M, Nagahama N, Toganoh M, Hackler U E, Isobe H and Nakamura E 2001 *Science* **291** 1944
[10] Patnaik A 2007 *J. Nanosci. Nanotechnol.* **7** 1111
[11] Nakashima N, Ishii T, Shirakusa M, Nakanishi T, Murakami H and Sagara T 2001 *Chem. Eur. J.* **7** 1766
[12] Nakanishi T, Schmitt W, Michinibu T, Kurth D G and Ariga K 2005 *Chem. Commun.* 5982
[13] Nakanishi T, Miyashita N, Michinobu T, Wakayama Y, Tsuruoka T, Ariga K and Kurth D G 2006 *J. Am. Chem. Soc.* **128** 6328
[14] Nakanishi T, Takahashi H, Michinobu T, Hill J P, Teranishi T and Ariga K 2008 *Thin Solid Films* **516** 2401
[15] Nakanishi T, Takahashi H, Michinobu T, Takeuchi M, Teranishi T and Ariga K 2008 *Colloids Surf. A* **321** 99
[16] Michinobu T, Nakanishi T, Hill J P, Funahashi M and Ariga K 2006 *J. Am. Chem. Soc.* **128** 10384
[17] Nakanishi T, Ariga K, Michinobu T, Yoshida K, Takahashi H, Teranishi T, Möhwald H and Kurth D G 2007 *Small* **3** 2019
[18] Nakanishi T, Wang J, Möhwald H, Kurth D G, Michinobu T, Takeuchi M and Ariga K 2009 *J. Nanosci. Nanotechnol.* **9** 550
[19] Zemb T, Dubois M, Demé B and Gulik-Krzywicki T 1999 *Science* **283** 816
[20] Yoshida K, Minamikawa H, Kamiya S and Shimizu T 2007 *J. Nanosci. Nanotechnol.* **7** 960
[21] Miyazawa K, Kuwasaki Y, Obayashi A and Kuwabara M 2002 *J. Mater. Res.* **17** 83
[22] Tan Z, Masuhara A, Kasai H, Nakanishi H and Oikawa H 2008 *Jpn. J. Appl. Phys.* **47** 1426
[23] Nakanishi T, Michinobu T, Yoshida K, Shirahata N, Ariga K, Möhwald H and Kurth D G 2008 *Adv. Mater.* **20** 443
[24] Sun T, Feng L, Gao X and Jiang L 2005 *Acc. Chem. Res.* **38** 644
[25] Frankevich E, Maruyama Y and Ogata H 1993 *Chem. Phys. Lett.* **214** 39
[26] Deschenaux R, Donnio B and Guillon D 2007 *New J. Chem.* **31** 1064
[27] Li W-S, Yamamoto Y, Fukushima T, Saeki A, Seki S, Tagawa S, Masunaga H, Sasaki S, Takata M and Aida T 2008 *J. Am. Chem. Soc.* **130** 8886
[28] Nakanishi T, Shen Y, Wang J, Yagai S, Funahashi M, Kato T, Fernandes P, Möhwald H and Kurth D G 2008 *J. Am. Chem. Soc.* **130** 9236

Journal of Physics: Conference Series **159** (2009) 012006 doi:10.1088/1742-6596/159/1/012006

Mechanical characterization of C_{60} whiskers by MEMS bend testing

MP Larsson and S Lucyszyn

Optical and Semiconductor Devices Group
Department of Electrical and Electronic Engineering, Imperial College London, Exhibition Road, London, SW7 2AZ, UK

E-mail : s.lucyszyn@imperial.ac.uk

Abstract. Little has been published on the mechanical characteristics of C_{60} whiskers, due to the inherent difficulties in physically mounting such small test samples. Earlier reported results suggested Young's modulus values of 32 and 54 GPa, with 130 and 160 nanometer diameter C_{60} whiskers, respectively, using compressive deformation techniques. In our work, an experimental bespoke silicon-based microelectromechanical system has been developed to extract another value. It has been found, through parameter extraction techniques, that a Young's modulus of only ~ 2 GPa is obtained with a C_{60} whisker having a diameter of 4 microns. By including the previously published data points, there is now strong evidence to suggest an inverse proportionality relationship between the Young's modulus and the diameter of a C_{60} whisker.

1. Introduction

Since the discovery of fullerene C_{60}, by Kroto *et al.* [1], advances in the field of fullerene chemistry have led to a number of interesting developments, such as the carbon nanotube (CNT). The C_{60} molecule consists of carbon atoms located at the nodes of a series of hexagons and pentagons arranged in a cage lattice, defined by alternating double and single bonds. In contrast to CNTs, which can be visualised as rolled-up sheets of graphene, C_{60} whiskers contain no long-range hollow structure. Instead, structural characterisation has indicated the whiskers are solid, consisting of a series of C_{60} molecules, bound through a combination of van der Waals interactions and chemical bonds. The latter suggests that C_{60} whiskers contain chains of polymerised C_{60} within their structure. A transition from a poly- to single-crystalline structure may occur when the diameter of C_{60} whiskers decreases to approximately 1 μm [2-3]. The absence of grain boundaries, combined with the observation of enhanced intermolecular bonding, are the main reasons why C_{60} nanowhiskers are expected to exhibit superior electrical and mechanical properties to those of their larger-diameter counterparts.

In recent years, work at the University of Tsukuba and National Institute for Material Science (both in Tsukuba, Japan), has led the way in determining the mechanical properties of C_{60} nanowhiskers using compressive deformation techniques [4]. Asaka *et al.* determined Young's modulus values of 32 and 54 GPa, with 130 and 160 nanometer diameter C_{60} whiskers, respectively. This is in excess of the 8.3 to 20 GPa range that they reported in their survey for bulk crystal C_{60} [4].

At Imperial College London, work on the electrical and mechanical characterization of C_{60} whiskers and nanowhiskers was undertaken within a two-year research programme [5-6]. Within this activity, a bespoke microelectromechanical systems (MEMS) electrothermal 4-point bend tester was developed for the mechanical characterization of C_{60} (nano)whiskers, for operation under an optical microscope. The characterization of Young's modulus, using the results from one of the experiments, will be investigated here.

2. Growth of C_{60} whiskers

C_{60} molecules are known to polymerise under high temperature [2-3], high pressure conditions and in the presence of ultraviolet (UV) radiation. Relatively recently, Miyazawa *et al.* reported a technique for growing needle-like structures of C_{60} from solution at room temperature. A logical first stage for our research was to reproduce the results [5] of Miyazawa *et al.* Here, the needles of C_{60} grew in arbitrary directions through a process of liquid-liquid interfacial precipitation (LLIP), between C_{60} saturated toluene and isopropyl alcohol (IPA), as shown in figure 1(a). Diameters ranged from hundreds of nanometres to several microns, with lengths reaching several hundreds of microns. Needles with diameters below 1 μm are referred to as C_{60} nanowhiskers, and those with larger diameters are described as C_{60} whiskers, both seen in figure 1(b). The DC characterisation of such samples has been recently reported [6].

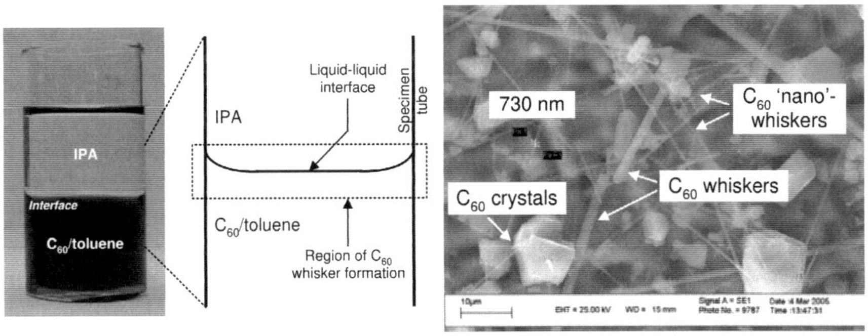

(a) (b)

Figure 1. Miyazawa's liquid-liquid interface growth technique: (a) growth vial indicating the region in which C_{60} whiskers are observed to grow in an ad hoc manner; and (b) products found within the liquid-liquid interface following several hours of growth. (Reproduced by permission of The Electrochemical Society. [6])

3. Mechanical bend testing

At the time of this work, no mechanical characterisation on C_{60} nanowhiskers had been reported in the open literature. However, relatively recently, the Young's modulus (E) of a C_{60} nanowhisker was ascertained from a buckling test in a transmission electron microscope (TEM), using a probe tip with position control and force-feedback [4]. Techniques for determining the mechanical properties of CNTs, in particular, have been widely reported in the open literature. The most common involves tensile loading using the tip of an atomic-force microscope (AFM) probe [7]. In such situations, the CNTs are either attached to the AFM tip by pressure contact or they are grown on. The former is more common; nevertheless, non-ideal attachment between CNT and the substrate and tip of the AFM limit the accuracy of the data obtained. The problem of gripping can be avoided by reversing the loading direction of the AFM tip and performing a buckling test. With this approach, the point of loading shifts during the test; once again invalidating the results. Problems with sample gripping or loading during mechanical testing can be resolved through the use of bend tests. Such tests are popular in mechanical engineering, for the characterisation of brittle materials. In such cases, 3- and 4-point tests are employed by convention, and the deflection and fracture data can be used to determine E and fracture strength; which for brittle materials is known as the modulus of rupture (MOR).

Micromachined tensile testers have successfully been developed for the characterisation of materials commonly used in MEMS technology, such as silicon and polysilicon. These devices are quite sophisticated, incorporating either electrothermal or electrostatic actuation. Force feedback can be achieved using electrostatic comb-drive sensors. It is the successful application of such devices that is the inspiration behind the mechanical testing of C_{60} (nano)whiskers. As C_{60} whiskers are large enough to be observed under an optical microscope, this lends itself well to the use of simplified MEMS mechanical testers, to evaluate their deflection and fracture characteristics.

3.1. MEMS bend tester

A MEMS electrothermal bend tester was designed for use under an optical microscope. The design was based around the dimensions and likely deflection range of a typical C_{60} (nano)whisker. Figure 2 illustrates the proposed device, containing a pair of back-to-back bend testers on a common die.

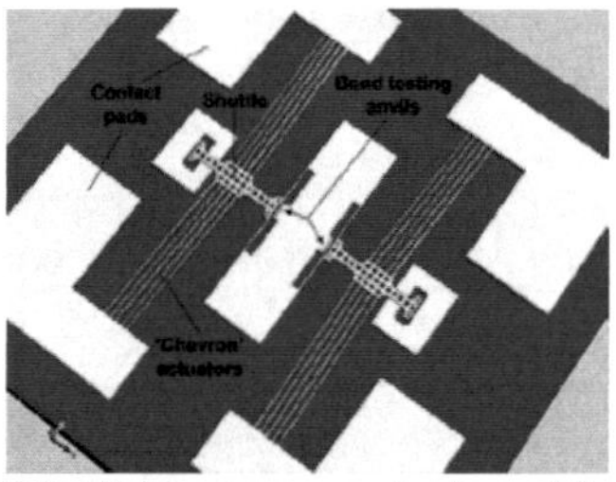

Figure 2. Proposed MEMS bend tester, designed for C_{60} (nano)whiskers.

Actuation is achieved by application of current between pairs of contact pads, causing heating within 'chevron' arms that supports a suspended 'shuttle' structure. The arms are angled in a manner to generate linear motion from their combined thermal expansion. The shuttle is attached to a set of anvils at one end, which are close to a fixed set of anvils (opposite), in order to apply bending point loads on a pre-positioned C_{60} whisker sample.

A process flow for the fabrication of the MEMS structure is shown in figure 3, relying on a two-stage deep reactive-ion etching (DRIE) of a bonded silicon-on-insulator (BSOI) wafer, to form an interdigitated set of (movable) anvil and (fixed) platform structures. Such features are important to support and raise the C_{60} (nano)whisker from the clearance zone between device and handle silicon layers, such that they can be loaded without falling into, and getting trapped within, the clearance area.

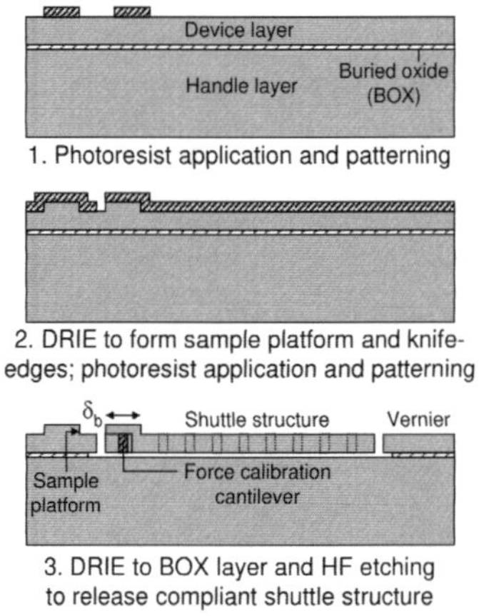

Figure 3. Process flow outlining key steps in the fabrication of the MEMS bend tester.

Prototypes were fabricated to test device performance and compare this with design predictions. Figure 4 shows optical microscope images of a prototype MEMS bend testers.

Journal of Physics: Conference Series **159** (2009) 012006 doi:10.1088/1742-6596/159/1/012006

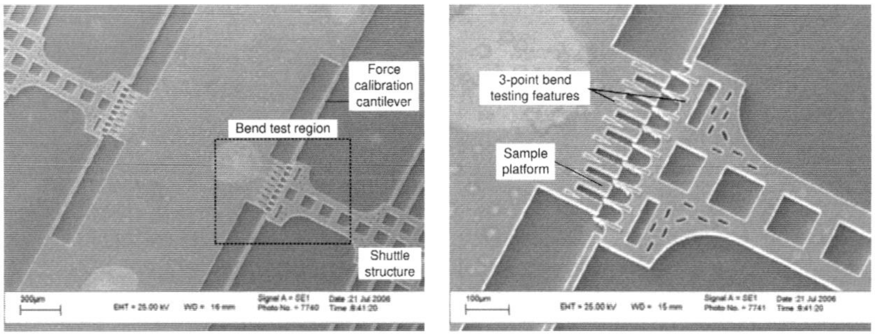

(a) (b)

Figure 4. Prototype MEMS bend testers: (a) central region showing both sets of loading anvils; and (b) detail over the region indicated in (a).

The MEMS bend tester was designed for ease of fabrication and use. To this end, features for rapid optical evaluation of force-deflection characteristics were included. Such features include a Vernier scale, with micron demarcations, and force calibration cantilevers to help in the determination of bending forces exerted on C_{60} (nano)whisker samples during testing. As the Young's modulus for silicon can be estimated to a good approximation, the force deflection characteristics for the cantilevers can be ascertained from their dimensions. Bend tests in the absence of samples will allow the deflection-force characteristics of the bend tester to be determined as a function of applied DC power. When a C_{60} (nano)whisker sample is inserted between the anvils, the force-deflection-power characteristic will change; the difference relative to the original being the additional force applied to the sample. This approach, although simple in theory and easy to realise in a MEMS device, can be difficult to achieve in practice. The problem lies in the ability to resolve differences in force-deflection-power characteristics, under an optical microscope, from normalised values after the insertion of a sample between the loading anvils. Even if quantitative data cannot be obtained, qualitative information on the mode of fracture and the extent of deformation prior to this would be useful information.

Bend testers with 5 and 10 supporting arms, either side of the central shuttle, were fabricated to vary the force applied to different samples. The spacing between opposite anvils was varied along each tester to accommodate whiskers of different diameters and lengths. Finally, testers with anvil arrangements to perform 3- and 4-point tests were designed. The latter allows more accurate data to be obtained, whereas the former generates greater observable deflections. The displacement performance of the prototype bend testers was evaluated by passing current in manner shown in figure 5(a). The thickness of the device layer in all prototypes was 10 μm, and resultant shuttle deflections-power densities are displayed in figure 5(b). For example, the graph shows that a 10-arm device produces a 12.5 μm/W deflection per Watt of applied DC power.

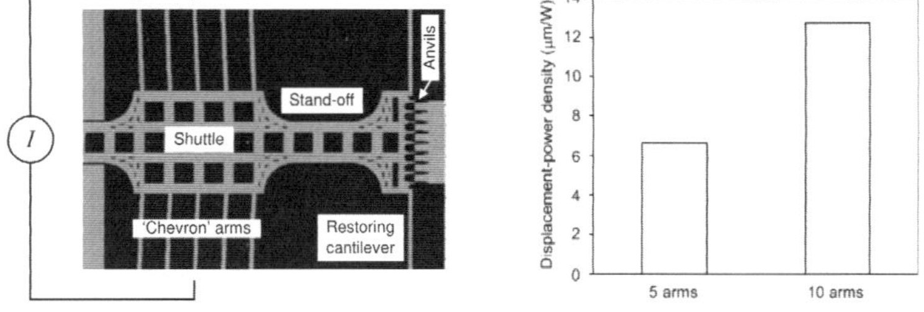

(a) (b)

Figure 5. (a) illustration showing the application of current to a tester; and (b) averaged deflection-power density data for devices with 5 and 10 arms on either side of the shuttle.

During performance evaluations, an infrared camera was used to image the thermal distribution throughout the device and measure its average temperature as a function of applied power. Figure 6 shows the variation in average device temperature with input power on a 10-arm bend tester. At full deflection, the device reaches a maximum temperature that is marginally in excess of 70 °C, corresponding to a power of approximately 2.7 W. Comparison between actual performance and design predictions revealed good agreement, showing the bend tester device to be functioning as intended. The next step involved the testing of real samples.

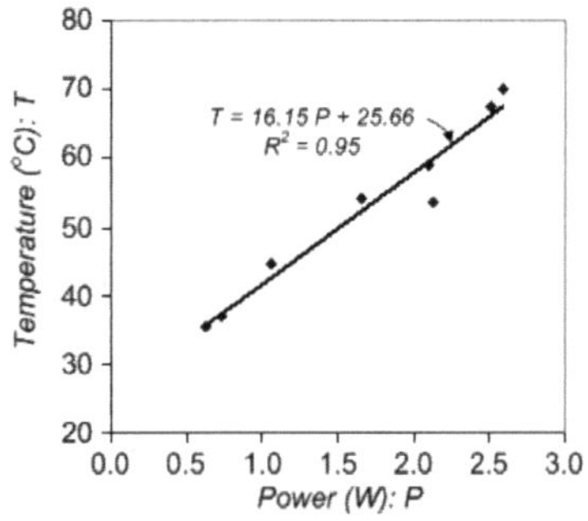

Figure 6. Variation in average die temperature with applied power.

3.2. Bend tests on C_{60} whiskers

The placement of C_{60} nanowhiskers between the anvils of the bend testers was always going to be difficult to achieve in practice. One approach adopted the direct application of a solvent-suspension of nanowhiskers over target areas. The disadvantage of this approach is that it is imprecise, and introduces 'debris' in the clearance zone between the movable shuttle and the substrate handle layer. Although the shuttle is capable of generating sufficient force to override such obstructions, the possibility of extracting *E* from a nanowhisker sample through differentials in force-displacement-power coefficients was not possible.

A more reliable approach for the precise positioning of C_{60} nanowhiskers in regions between the anvils of the bend testers was eventually employed. The technique involved the use of a flame-stretched glass capillary tube, having a tip diameter of 5 μm. By attaching the capillary tube to a Mitutoyo *x*-*y*-*z*-positioner, a high-resolution positioning device could be realised, as shown in figure 7. If two are used in parallel, C_{60} nanowhiskers can be moved into position with considerable ease.

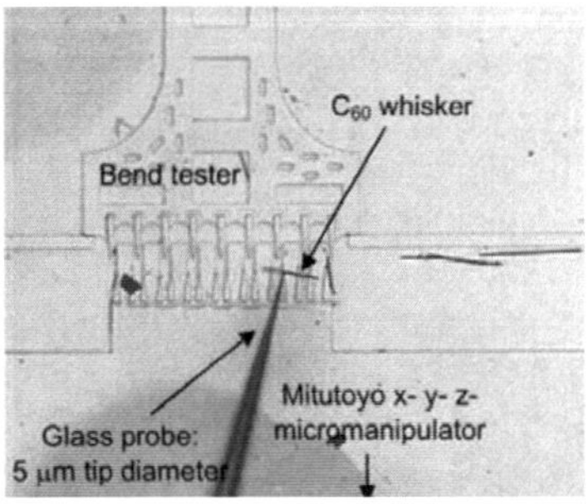

Figure 7. Placement of a C_{60} whisker between the anvils of a bend tester, using a micropositioner attached to a capillary tube having a 5 μm tip diameter.

In practice, the glass capillary tubes used were coated in a sputtered layer of Au, to maintain a clean contact surface and promote electrostatic attraction with nanowhiskers. It is known that Au surfaces give up electrons during rubbing contact, due to the triboelectric effect. This will allowed the capillary tubes to establish a positive static charge of sufficient magnitude to attract nanowhiskers, which have been shown to develop an attraction to positively charged surfaces. In some cases, this 'pick-n-place' approach did not work, as van der Waal's forces between nanowhiskers and the silicon surface were too great. Such surface interactions tended to be greater with the narrowest nanowhiskers. In some cases, the nanowhiskers could be dislodged from the silicon surface using the probe tip; however, the difference in size between the two often caused the nanowhisker to fracture. As such, the narrowest nanowhiskers could not be successfully positioned using this technique, and only samples with diameters in excess of 0.5 μm could be placed between the anvils of the bend testers.

Optical microscope images of samples in 3- and 4-point bend testers are shown in figures 8 and 9, respectively. A fractured C_{60} nanowhisker is seen in figure 8(a), following a prior successful 3-point bend test. Failure appears to be entirely brittle, as no signs of plastic deformation are visible. Figure 8(b) shows the same tester during the application of DC power. The larger C_{60} whisker, to the right of the C_{60} nanowhisker, is now experiencing loading between the three contacting anvils. Repeated loading and unloading revealed elastic deformation characteristics. Unfortunately, in the process of this basic 'fatigue' test, the sample worked its way above the anvils and could not be tested to destruction.

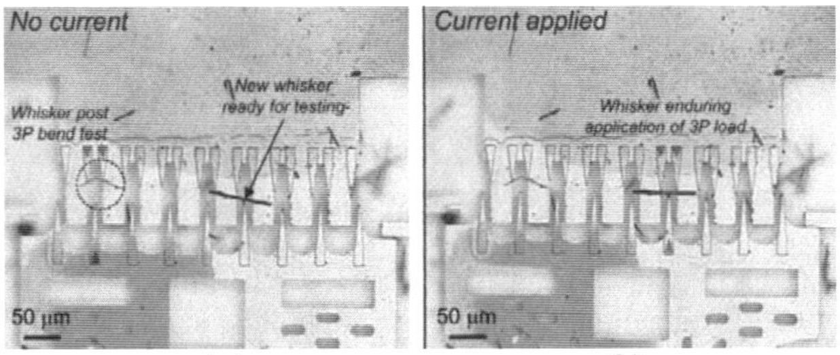

(a) (b)

Figure 8. 3-point bend test of C_{60} whiskers: (a) before; and (b) during tester actuation.

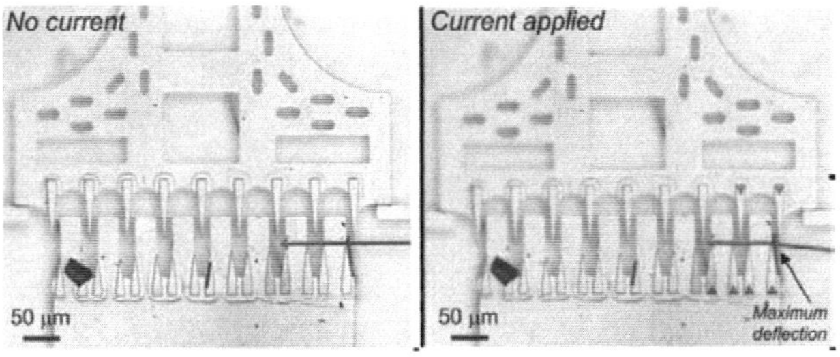

(a) (b)

Figure 9. 4-point bend test of a C_{60} whisker: (a) before; and (b) during tester actuation.

Closer inspection of the image in figure 9(b) reveals that the loading anvil at the far right of the tester is marginally longer than the other three anvils in contact with the whisker. So despite the fact that the whisker is clearly under load from all four anvils, the greatest deformation (and, hence stress) is occurring in the region above the third from the left anvil. The unusual situation conveniently produces what is in effect a cantilever bend test, with the 'fixed' support provided by anvils 1-3 (left to right), and the point load applied by the 4th anvil. Figure 10 shows a simplified schematic representation of the cantilever bend test between the 3rd and 4th anvils.

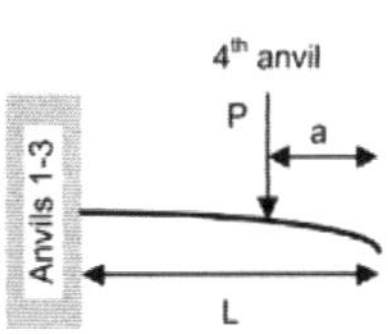

Figure 10. Cantilever bending due to a transverse point-load, representing a possible scenario for loading between anvils 3 and 4 of the bend tester in figure 9.

The governing equation that relates the deflection of the cantilever's tip is given by [8]:

$$\delta = \frac{P}{6EI}\left(2L^3 - 3L^2 a + a^3\right) \tag{1}$$

where δ is the tip deflection, I is the second moment of area of the cantilever, and P is the applied load. The image in figure 9(b) was captured at the instant immediately prior to sample fracture, which occurred at the 4th anvil. The deflection observed is the maximum for the sample prior to fracture. The distance between the 3rd and 4th anvils is approximately 10 μm, the maximum deflection noted at the point of load application was measured at just under 1.5 μm and the whisker's diameter was determined to be approximately 4 μm. Figure 11 shows the fractured whisker following successful bend testing. The fracture point indicates brittle failure.

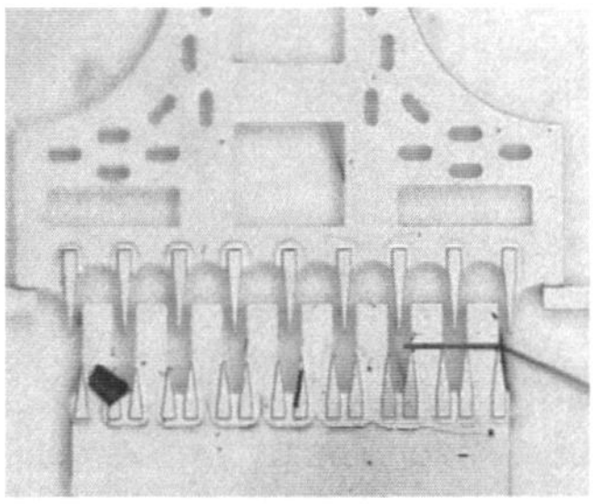

Figure 11. C_{60} whisker showing signs of brittle fracture following a bend test.

3.3. Experimental results

The resolution of the optical microscope and deflection measurement system used were not sufficient to allow representative load force data to be inferred from applied DC power, as initially envisaged. An approximation for the Young's modulus of the C_{60} whisker in figure 9 is given, assuming a maximum resolution for the optical system of 1 μm. If there is no change observed in the force-deflection-power characteristics of the bend tester, following sample insertion, the additional load force applied to the sample corresponds to that required to move the unloaded tester through a distance of ~ 1 μm. Based on the dimensions of the force calibration cantilevers and an assumed E for silicon of 160 GPa, this corresponds to a force of ~ 2.5 μN; in other words, the bend tester has a force-deflection characteristic of 2.5 μN/μm. Assuming this force to be of the same order of magnitude to the maximum load applied to the whisker in the bend test, a simple calculation using equation (1) yields an approximation of its stiffness and, hence, E. By assuming the whisker's shape to follow that of a solid rod of diameter, D, the second moment of area, I, is calculated using the following standard expression:

$$I = \frac{\pi D^4}{64} \tag{2}$$

Although the whisker's tip deflection is cropped from the image in figure 9(b), its value was measured, and with remaining measurements on the loaded sample inserted into equation (1), to yield a value of E ~ 2 GPa. This value is at least a factor of 10 lower than that reported by Asaka *et al.* [4], with their axially-loaded C_{60} nanowhisker having a diameter of 160 nm. Our C_{60} whisker has a diameter of 4 μm, which is at least 25 times thicker than those reported by Asaka *et al.* [4]. This makes it an interesting sample for comparison. From the limited experimental data available (two data points from Asaka *et al.* and one from our work), Young's modulus can be crudely modelled, as shown in figure 12.

Ideally, more data points from controlled samples are needed in order to determine a more definitive model, but the inverse proportionality of Young's modulus to C_{60} whiskers diameter is evident. This appears to contradict any notion of a fixed value of Young's modulus for C_{60} whiskers.

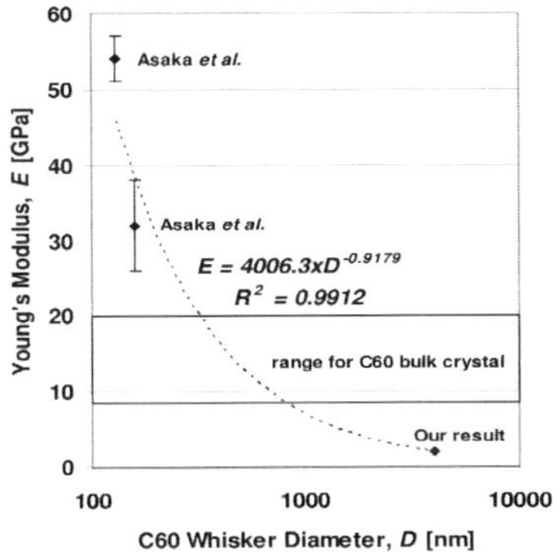

Figure 12. Characterization of Young's modulus with C_{60} whisker diameter.

4. Discussion and conclusions

The lower stiffness of the whisker characterised in this study is likely to be a consequence of the presence of grain boundaries due to its polycrystalline structure. Of course, another possibility is that the technique employed leads to an underestimate of the stiffness of the whisker. One possible source of error is the measurement of maximum whisker deflection prior to failure. Although previous deflections can equally be used to determine E, the final value was used as this facilitated measurement through the optical microscope. Other, more minor, errors may come from the approximation leading to the scenario represented by figure 10 and subsequent use of equation (1).

Despite the limitations of the mechanical characterisation carried out in this study, the results are informative on qualitative grounds as they provide evidence for brittle fracture in C_{60} nanowhiskers as well as larger diameter whiskers. Our parameter extraction method suggests a value for $E \sim 2$ GPa for a 4 μm diameter C_{60} whisker. By including the previously published data points, there is strong evidence to suggest an inverse proportionality relationship between the Young's modulus and the diameter of the C_{60} whisker.

Acknowledgements

The authors are very grateful to Dr Kun'ichi Miyazawa, for providing the initial inspiration for our work on C_{60} whiskers and ongoing discussions. We also like to thank the UK's Engineering and Physical Research Council for funding this research (under grant GR/S97019/01).

References

[1] Kroto H W, Heath J R, O'Brien S C, Curl R F and Smalley R E 1985 C_{60}: buckminsterfullerene *Nature* **318** 162-3

[2] Miyazawa K, Kuwasaki Y, Obayashi A and Kuwabara M 2002 C_{60} nanowhiskers formed by the liquid-liquid interfacial precipitation method *J. of Mat. Res.* **17** 83-8

[3] Miyazawa K, Kuwasaki Y, Hamamoto K, Nagata S, Obayashi A and Kuwabara M 2003 Structural characterization of C_{60} nanowhiskers formed by the liquid/liquid interfacial precipitation method *Surf. and Int. Ana.* **35** 117-20

[4] Asaka K, Kato R., Miyazawa K, Kizuka T 2006 Buckling of C_{60} whiskers *App Phys. Lett.* **89** 071912

[5] Larsson M P and Lucyszyn S 2005 Directed growth of C_{60} nanowhiskers for millimetre-wave detectors *Romanian J. of Inf. Sci. and Tech.* **8** 305-20

[6] Larsson M P, Kjelstrup-Hansen J and Lucyszyn S 2007 Dc characterisation of C_{60} whiskers and nanowhiskers *ECS Trans.* **2** 27-38

[7] Yu M F, Files B S, Arepalli S and Ruoff R S 2000 Tensile loading of ropes of single wall carbon nanotubes and their mechanical properties *Phy. Rev. Lett.* **84** 5552-5

[8] Young W C and Budynas R G 2002 Roark's formulas for stress and strain (McGraw-Hill International Edition)

Fullerene nanowhiskers and related fullerene nanomaterials

K Miyazawa[1], R Kato[1], K Saito[2], T Kizuka[2], T Mashino[3] and S Nakamura[3]

[1]National Institute for Materials Science, Tsukuba, 305-0044, Japan
[2]Institute of Materials Science, University of Tsukuba, Tsukuba 305-8573, Japan
[3]Faculty of Pharmacy, Keio University, Minato-ku, Tokyo, 105-8512, Japan

E-mail : miyazawa.kunichi@nims.go.jp

Abstract. Since the discovery of C_{60} nanowhiskers in a colloidal solution of PZT in 2001, various low-dimensional fullerene nanomaterials such as fullerene nanotubes and fullerene nanosheets have been synthesized by the liquid-liquid interfacial precipitation method (LLIP method) as well as the C_{60} and C_{70} nanowhiskers. The LLIP method is an excellent process to synthesize the quasi one-dimensional fullerene nanomaterials. This paper first reviews the typical fullerene nano and micro materials that have been prepared by the LLIP method and then show that the LLIP method is also applicable to the synthesis of composite C_{60} nano and micro whiskers containing wide compositions of C_{60} derivative molecules.

1. Introduction

A variety of novel low-dimensional fullerene nanomaterials have been synthesized since the discovery of C_{60} nanowhiskers (C_{60}NWs) in a colloidal solution of lead zirconate titanate (PZT) added with C_{60} in 2001 [1]. The C_{60}NWs can be classified into the category of "fullerene nanofibers (FNFs)". The FNFs are the thin fibers with diameters less than 1000 nm and are composed of various fullerene molecules that include endohedral fullerenes and fullerene derivative molecules with functional groups. The FNFs have the aspect ratios of length/diameter greater than three and can take single crystalline, polycrystalline or amorphous structures. The FNF can have a very wide range of length from a few micrometers to several millimetres or more. The FNFs can contain impurity elements and solvents in their matrices as well.

The fullerene nanowhiskers (FNWs) are the needle-like crystals that are composed of fullerene molecules and have diameters less than 1000 nm, and hence constitute a part of FNFs. On the other hand, the fullerene nanotubes (FNTs) are the FNFs with a tubular structure [7], and can take single crystalline, polycrystalline or amorphous structures. Fullerene nanorods and fullerene nanowires are classified into the category of FNFs. Recently, short FNWs are sometimes called "fullerene nanorods" [2, 3, 4] and the fullerene nanofibers are occasionally called "fullerene nanowires" [5].

One of the most efficient and widely used methods to prepare FNFs is the liquid-liquid interfacial precipitation method (LLIP method) that can even produce single crystalline fullerene nanotubes and fullerene nanosheets as well as the non tubular FNWs [6,7,8,9].

In general, the heterogeneous nucleation of crystals more easily occurs than the homogeneous nucleation in solutions. Substrates are used for such a purpose to enhance the nucleation of crystals. The interface formed between two liquid phases also can be used for the above purpose. The LLIP method produces the crystal nuclei of fullerenes by laying an alcohol on a fullerene-saturated solution from which FNFs grow. The LLIP method was firstly applied for the C_{60}NWs prepared by use of a

C_{60}-saturated toluene solution and isopropyl alcohol (IPA) [6]. A supersaturation state for C_{60} is realized by adding IPA to the toluene solution of C_{60} owing to the interdiffusion between toluene and IPA. The crystal nuclei produced at the liquid-liquid interface grow and C_{60}NWs appear. The alcohols act as the nucleators of fullerene nanocrystals. By changing the combination of solvents, tubular fullerene nanofibers can be obtained. Single crystalline nanotubes composed of C_{60} or C_{70} can be fabricated by using their pyridine solutions and IPA [7, 8].

The LLIP method can also be applied to prepare "fullerene nanosheets", single crystalline thin sheets made of C_{60} [9]. A variety sizes of C_{60} nanosheets have been synthesized by a modified LLIP method, using a C_{60}-CCl_4 solution and alcohols (isopropyl alcohol, ethanol, methanol). The diameter of C_{60} nanosheets changes depending on the kind of alcohols from 7.5 μm (CCl_4/IPA) to 500 nm (CCl_4/methanol). The self-standing single crystalline C_{60} nanosheets may be used as organic semiconductor films. The LLIP method is also applicable to the synthesis of various crystals other than fullerene crystals. For example, asymmetric NaCl crystals can be synthesized by the LLIP method using the interface formed between a water solution of NaCl and 1-butanol [10].

Furthermore, the LLIP method has been developed to "the liquid-membrane-liquid interfacial precipitation (LMLIP) method", where an alcohol is injected into a fullerene-saturated solution through a membrane with fine through-holes. Cha *et al.*, showed that vertically aligned microtubes of C_{60} can be fabricated by injecting IPA into a toluene solution of C_{60} through a porous alumina membrane [11]. The surface of alumina membranes supplies the nucleation sites of C_{60} from which the C_{60} microtubes grow.

Not only the monocomponent FNFs but also multicomponent FNFs can be prepared by use of LLIP method. The first example was the two - component fullerene nanotubes with a composition of C_{60} - 15 mol% C_{70} that were prepared by use of IPA and a pyridine solution saturated with a powder mixture of C_{60} and C_{70} [7].

The properties of fullerene nanofibers are expected to be variously modified by adding fullerene derivative molecules to the matrix of fullerene nanofibers as well. The first synthesis of FNFs containing fullerene derivatives was the C_{60} nanowhiskers added with $(\eta^2\text{-}C_{60})Pt(PPh_3)_2$ molecules [12,13]. Owing to the large size of $(\eta^2\text{-}C_{60})Pt(PPh_3)_2$, C_{60} powders containing more than 1 mol% $(\eta^2\text{-}C_{60})Pt(PPh_3)_2$ could not produce the C_{60} nanowhiskers, while capsular or tubular C_{60} needle-like crystals appeared. This result shows an existence of a solubility limit of C_{60} derivative molecules to the matrix of C_{60} nanowhiskers. Hence, it is of importance to prepare the C_{60} nanowhiskers containing various C_{60} derivatives with various compositions in order to examine the extent and possibility of LLIP method. The present paper will show Raman spectroscopic analyses of C_{60} nanowhiskers containing four kinds of C_{60} derivatives and discusses their structural characteristics deduced from the Raman spectroscopy data.

2. Experimental

Various two-component fullerene whiskers containing the C_{60} derivatives as shown in Fig.1 were prepared by the LLIP method.

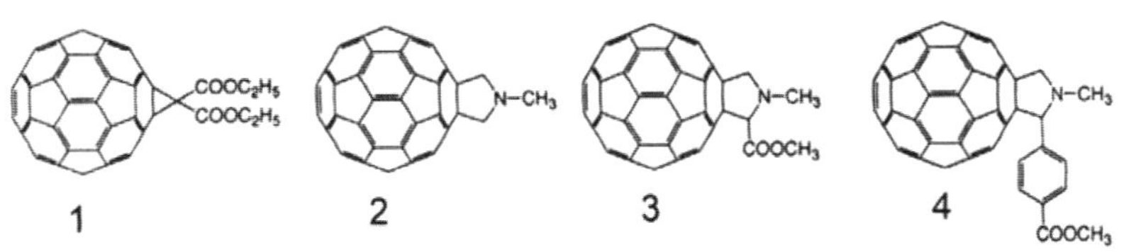

Fig. 1 C_{60} derivatives used for the preparation of two-component C_{60} whiskers.

The LLIP process was done by using *m*-xylene for the good solvent of C_{60} and the C_{60} derivatives and IPA for the poor solvent. The high-performance liquid chromatography (HPLC) analysis was done by dissolving the C_{60} whiskers into *m*-xylene. Fractions were separated on an octadecylsilyl

(ODS) column (150×4.6 mm). The elution was performed using a mobile phase of ethyl acetate : toluene = 9:1 at a flow rate of 0.5 ml min^{-1}. The prepared fullerene whiskers were observed by an optical microscope and also examined by Raman spectroscopy at a laser excitation wavelength of 532 nm (JASCO, NRS-3100, Japan).

3. Results and discussion

The prepared fullerene whiskers were found to have the compositions smaller than those of the starting powders as shown in Fig.2. The compositions analyzed by HPLC deviate from the nominal starting compositions with increasing those values. Comparing the C_{60} derivatives 2 and 3, the deviation from the stoichiometry line is larger for the C_{60} derivative 3 with a larger substituent group, which suggests that the dissolution of C_{60} derivative molecules into the matrix of C_{60} fullerene whiskers becomes more difficult with increasing the substituent size.

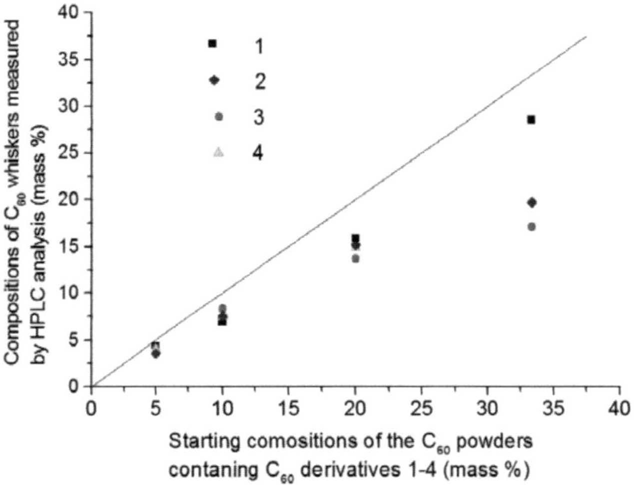

Fig. 2 Compositions of the C_{60} whiskers prepared by use of the C_{60} powders containing C_{60} erivatives 1 - 4.

As shown in Figs. 3 and 4, various morphologies of fullerene crystals were obtained by the LLIP method using the C_{60} powders containing the C_{60} derivatives 1 - 4. Thin fullerene nanowhiskers are mostly observed in Fig.3 (a), Fig. 3(b) and Fig. 4(a), while thicker needle-like crystals or the precipitates with irregular shapes are shown in Fig. 3(c), Fig. 3(d), Fig. 4(b), Fig. 4(c) and Fig. 4(d). This result shows that the addition of higher concentration of C_{60} derivatives with larger sizes suppresses the growth of FNWs in general. This fact is consistent with our previous result of the preparation of C_{60} nanowhiskers containing $(\eta^2\text{-}C_{60})Pt(PPh_3)_2$ [12].

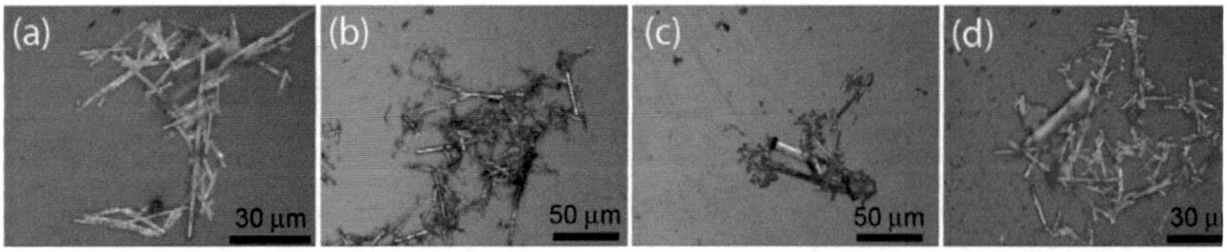

Fig.3 Optical micrographs of the fullerene whiskers and precipitates with the compositions, (a) C_{60}- 4.3 mass % C_{60} derivative 1, (b) C_{60} - 3.5 mass % C_{60} derivative 2, (c) C_{60} - 4.1 mass % C_{60} derivative 3, and (d) C_{60} - 4.0 mass % C_{60} derivative 4.

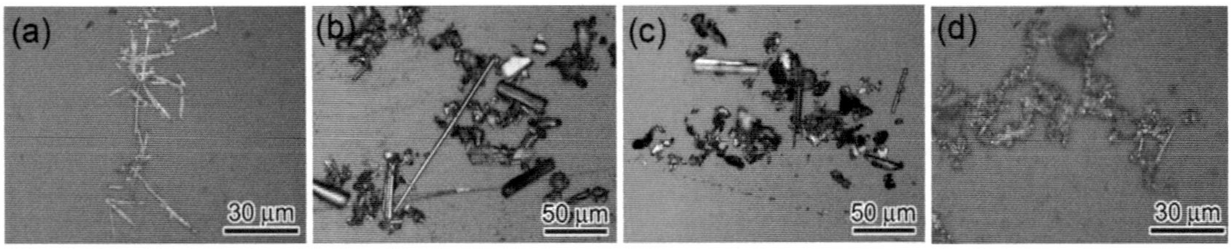

Fig. 4 Optical micrographs of the fullerene whiskers and precipitates with the compositions, (a) C_{60}-29 mass % C_{60} derivative 1, (b)C_{60} - 20 mass % C_{60} derivative 2, (c) C_{60} - 17 mass % C_{60} derivative 3, and (d) C_{60} -15 mass % C_{60} derivative 4.

The Raman spectroscopy is a powerful tool to study the chemical bonding states of C_{60}. Pure C_{60} molecules excite two A_g peaks and eight H_g peaks [14]. The $A_g(2)$ peak is sensitive to the deviation from the I_h symmetry of C_{60} and locates in the vicinity of 1469 cm^{-1} in pristine C_{60}. Fig. 5 compares the $A_g(2)$ peaks of as-prepared C_{60}NWs and those of the C_{60}NWs heated at various temperatures in vacuum to remove the contained organic solvents. The C_{60}NWs were prepared by use of a C_{60}-saturated pyridine solution and isopropyl alcohol [7]. The $A_g(2)$ peaks of the as-prepared C_{60}NWs show a Raman shift of ~1460 cm^{-1}, while the heated C_{60} NWs show the $A_g(2)$ peaks close to pristine C_{60}. Although the downward shift of $A_g(2)$ peak to 1460 cm^{-1} indicates a polymerization of C_{60} molecules [15], the origin of peak shift may be also explained by an interaction between the C_{60} molecules and the solvent molecules contained in the as-prepared C_{60}NWs. It is suggested that C_{60} molecules of the as-prepared C_{60}NWs are deformed from their pristine I_h structure owing to their interaction with the contained solvent molecules.

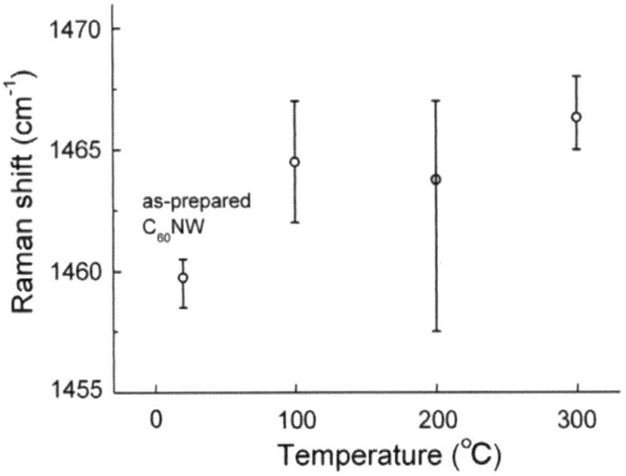

Fig.5 Wave numbers of $A_g(2)$ peaks of the as-prepared and heat-treated C_{60} nanowhiskers.

As shown in Fig. 6, every C_{60} derivative powder shows the split $H_g(1)$ peaks at around 255 cm^{-1} that are excited by the C_{60} cages deformed from the I_h symmetry of pristine C_{60}. However, those extra peaks are very weak or cannot be detected in the composite fullerene whiskers as shown in Fig. 7. The $A_g(2)$ peak is also sensitive to the deviation from the I_h symmetry, and is close to 1460 cm^{-1} in each profile of Fig. 6. However, like the case of $H_g(1)$ peaks, the $A_g(2)$ peaks are at around 1467 cm^{-1} in Fig. 7. This result shows that the solid-soluted C_{60} derivative molecules are weakly bound to the surrounding C_{60} molecules and that the C_{60} cages in the composite whiskers are also weakly bound by van der Waals bonding forces like the case of pristine C_{60} crystals. These observations suggest that the C_{60} derivative molecules dissolved into the matrices of C_{60} whiskers do not significantly alter their mechanical properties in a wide range of compositions and that the above C_{60} - C_{60} derivative composite whiskers have the Young's moduli similar to pristine C_{60}NWs. The Young's modulus of C_{60}NWs ranges from 32 - 54 GPa, which corresponds to 160 – 650 % of C_{60} bulk crystals [16].

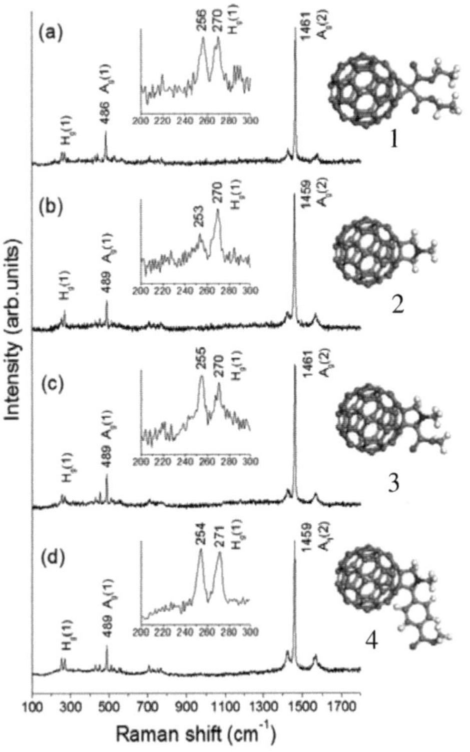

Fig.6 Raman profiles of the powders of C_{60} derivatives 1 - 4 of Fig.2.

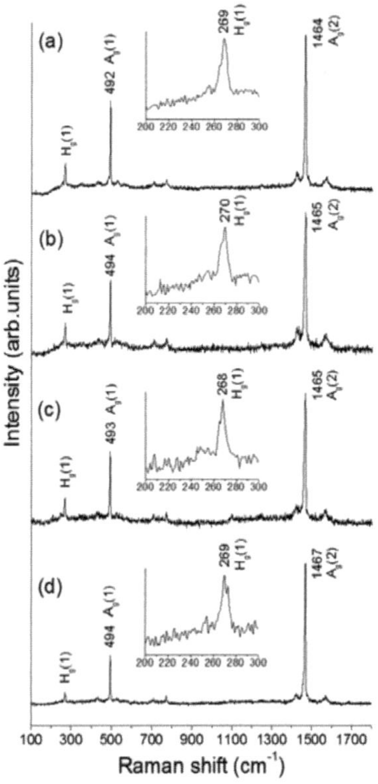

Fig.7 Raman profiles for the C_{60} whiskers containing C_{60} derivative molecules. (a) C_{60}-29 mass% C_{60} derivative 1, (b)C_{60} - 20 mass% C_{60} derivative 2, (c) C_{60} - 17 mass% C_{60} derivative 3, and (d) C_{60} -15 mass% C_{60} derivative 4.

4. Conclusions

The LLIP method is an excellent technique to prepare various morphologies of low-dimensional fullerene nanomaterials like fullerene nanowhiskers, fullerene nanotubes, fullerene nanosheets and vertically aligned fullerene microtubes at room temperature.

The LLIP method can produce also the multicomponent FNWs containing various C_{60} derivatives. However, the growth of FNWs is found to be inhibited by the addition of higher concentration of C_{60} derivatives with larger substituents.

The functional groups of the C_{60} cages located on the surface of FNWs will be further modified by attaching a variety of molecules like dendrimers. Hence, numerous kinds of quasi one-dimensional fullerene nanomaterials are expected to be produced in future. The low-dimensional fullerene nanomaterials are now establishing a new big scientific field of carbon.

References

[1] Miyazawa K Obayashi A and Kuwabara M 2001 *J.Am.Ceram.Soc.* **84** 3037

[2] Ji H, Hu J, Wan L, Tang Q and Hu W 2008 *J. Mater. Chem.* **18** 328

[3] Hou Y, Liu B, Wang L, Yu S, Yao M, Chen A, Liu D, Zou Y, Li Z, Zou B, Cui T, Zou G, Iwasiewicz-Wabnig A and Sundqvist B 2007 *J. Phys.: Condens.Matter* **19** 425207

[4] Malik S, Fujita N, Mukhopadhyay P, Goto Y, Kaneko K, Ikeda T and Shinkai S 2007 *J. Mater. Chem.* **17** 2454

[5] Geng J, Zhou W, Skelton P, Yue W, Kinloch I.A.,Windle A.H. and Johnson B.F.G. 2008 *J. Am. Chem. Soc.* **130** 2527

[6] Miyazawa K, Kuwasaki Y, Obayashi A and Kuwabara M 2002 *J.Mater.Res.* **17** 83

[7] Miyazawa K, Minato J, Yoshii T, Fujino M and Suga T 2005 *J.Mater.Res.* **20** 688

[8] Minato J and Miyazawa K 2006 *Diamond & Related Materials* **15** 1151

[9] Sathish M and Miyazawa K 2007 *J.Am Chem Soc.* **129** 13816

[10] Kadota K, Tanida S, Shirakawa Y, Shimosaka A and Hidaka J 2007 *J. Chem. Eng. Jpn* **40** 217

[11] Cha S I, Miyazawa K and Kim J D 2008 *Chem.Mater.* **20** 1667

[12] Miyazawa K and Suga T 2004 *J.Mater.Res.* **19** 2410

[13] Miyazawa K and Suga T 2004 *J.Mater.Res.* **19** 3145

[14] Weeks D E and Harter W G 1989 *Journal of Chemical Physics* **90** 4744

[15] Rao A M, Zhou P, Wang K A, Hager G T, Holden J M,Wang Y, Lee W T, Bi X X, Eklund P C, Cornet D S, Duncan M A and Amster I J 1993 *Science* **259** 955

[16] Asaka A, Kato R, Miyazawa K and Kizuka T 2006 *Appl. Phys. Lett.* **89** 071912

Observation of phagocytosis of fullerene nanowhiskers by PMA-treated THP-1 cells

S Nudejima[1], K Miyazawa[1], J Okuda-Shimazaki[2] and A Taniguchi[2]
[1] Advanced Nano Materials Laboratory, National Institute for Materials Science (NIMS), Tsukuba, Ibaraki, 305-0044, Japan
[2] Biomaterials Center, NIMS, Tsukuba, Ibaraki, 305-0044, Japan

E-mail: NUDEJIMA.Shinichi@nims.go.jp

Abstract. Phorbol 12-myristate 13-acetate (PMA)-treated THP-1 cells (macrophage-like cells) were exposed to the C_{60} fullerene nanowhiskers (C_{60}NWs) with an average length of about 6.0 μm and an average diameter of about 660 nm and observed with an inverted optical phase-contrast microscope for 48 h. The C_{60}NWs were well and stably dispersed onto the dishes of culture medium during the observation. The number of cells that internalised C_{60}NWs gradually increased after the exposure to C_{60}NWs. But no alteration of cellular morphology was observed compared to the control group without exposure to C_{60}NWs during this period in this pilot study.

1. Introduction

Nanomaterials possess enormous potentials to wide applications in various fields owing to their distinctive unique properties. But potential risks caused by exposure to nanomaterials have not been cleared. With the release of industrial products of nanomaterials into environment, public concerns have raised their potential side effects. Carbon nanotubes (CNTs), one of the most promising nanomaterials, may be hazardous to health and environment owing to their needle-like morphology and strong mechanical properties like asbestos. Recently, an asbestos-like pathogenic behavior associated with CNTs indicated a structure-activity relationship based on length, to which asbestos and other pathogenic fibers conform [1].

Fullerene nanowhiskers (FNWs) are composed of the fullerene molecules that are usually bonded via van der Waals forces [2]. FNWs are expected to have various application fields such as low-dimensional semiconductors, field emission tips, nanoprobes for microdevices, fiber-reinforced nanocomposites, composite elements for lubrication, and so on. But FNWs also have the needle-like morphology like asbestos. Hence, it is of great importance to evaluate the biological impacts of FNWs for their sound application in advance.

Macrophages are one of the immune system cells and defend the host against the foreign substances in a non-specific manner during the initial phase of infection. Macrophages recognize, internalise and digest them. THP-1 is a human acute monocytic leukemia cell line and THP-1 cells have been isolated by Tsuchiya et al [3]. It is well known that THP-1 cells are induced to differentiate into macrophage-like cells by the addition of PMA [4].

In this paper, the interaction of C_{60}NWs with macrophage-like cells is investigated as a pilot study for evaluating the biological impacts of C_{60}NWs by use of an inverted optical phase-contrast microscope.

2. Materials and methods

2.1. Preparation and characterization of C_{60}NWs

C_{60}NWs were prepared by the liquid-liquid interfacial precipitation method using a C_{60}-saturated toluene solution and isopropyl alcohol (Fig.1) [2]. 20 mL of isopropyl alcohol was gently added to 20 mL of C_{60}-saturated toluene solution in a glass bottle at room temperature. After manual mixing, the solution was kept at 15°C for 15 minutes and 60 mL of isopropyl alcohol was poured into the solution to stop the crystal growth. The C_{60}NWs were separated by filtration using a GFP filter (0.8 μm, Kiriyama glass, Japan) from the solvents. The characterization was carried out by measuring the length and diameter of C_{60}NWs using an optical microscope (ECLIPSE ME600, Nikon, Japan) and a scanning electron microscope (JSM-6700, JEOL, Japan). C_{60} (99.5%) was purchased from MTR (OH). Toluene (99.5%) and 2-propanol (99.7%) were purchased from Wako Pure Chemical Industries (Japan).

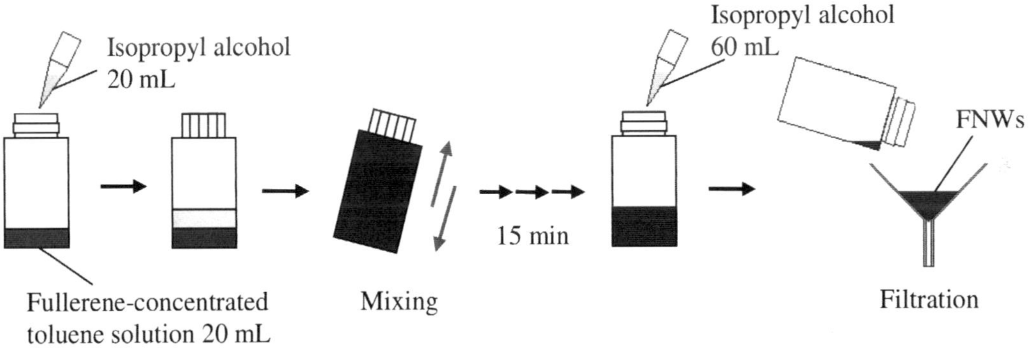

Figure 1. Preparation of C_{60}NWs

2.2. Cell culture

THP-1 cells were purchased from American Type Culture Collection (ATCC, VA). THP-1 cells were cultured in RPMI 1640 medium (Invitrogen, CA) supplemented with 10% heat inactivated fetal bovine serum (JRH Biosciences, KS), 100 units/mL penicillin and 100 μg/mL streptomycin (Nacalai Tesque, Japan)(culture solution) at 37°C in an atmosphere of 5% CO_2 and saturated humidity. Cells were subcultured every three or four days, where the number of cells in culture were maintained by centrifugation (at 1000 rpm for 3 min) and subsequent resuspension at 2 x 10^5 viable cells/mL.

2.3. Differentiation of THP-1 cells into macrophage-like cells

2 x 10^5 cells were incubated in 2 mL of culture solution on a 35 mm polystyrene culture dish (Greiner Bio-One, Germany). PMA (Wako pure chemicals, Japan) was dissolved in dimethylsulfoxide at a concentration of 1 mM and diluted by a culture solution to be 50, 500 and 5000 nM. And 40 μL of each PMA solution was added to the cellular medium to be the final concentrations of 1, 10 and 100 nM. The cells were induced to differentiate into macrophage-like cells for 1, 3, 6, 24, 48 and 72 h at 37°C in an atmosphere of 5% CO_2 and saturated humidity. To estimate the degree of differentiation, the number of suspended living cells (undifferentiated cells) was measured using 0.4% trypan blue stain (Invitrogen, CA) and the morphological changes of cells were observed by an inverted optical phase-contrast microscope (DMIL-HC, Leica Microsystems, Germany).

2.4. C_{60}NWs' exposure

C_{60}NWs were dispersed in the culture solution at a concentration of 1 mg/mL and diluted by the culture solution to be 0.1 and 0.01 mg/mL. Macrophage-like cells were exposed to 20 μL of each concentration of C_{60}NWs suspension agitated by ultrasonication to be the final concentrations of 0.1, 1 and 10 μg/mL and incubated for 1, 3, 6, 24 and 48 h at 37°C in an atmosphere of 5% CO_2 and saturated humidity. Cells were observed with the inverted optical phase-contrast microscope to evaluate the effect of C_{60}NWs on the cell morphology.

3. Results and disucussion

3.1. C_{60}NWs

3.1.1. Characterization of C_{60}NWs
C_{60}NWs were synthesized by the liquid-liquid interfacial precipitation method (Fig.2)[2]. The length of C_{60}NWs ranged from 1 to 17 μm approximately and the average length was about 6.0 μm (Fig.3). The diameter of C_{60}NWs ranged from 300 to 1340 nm and the average diameter was about 660 nm (Fig.4).

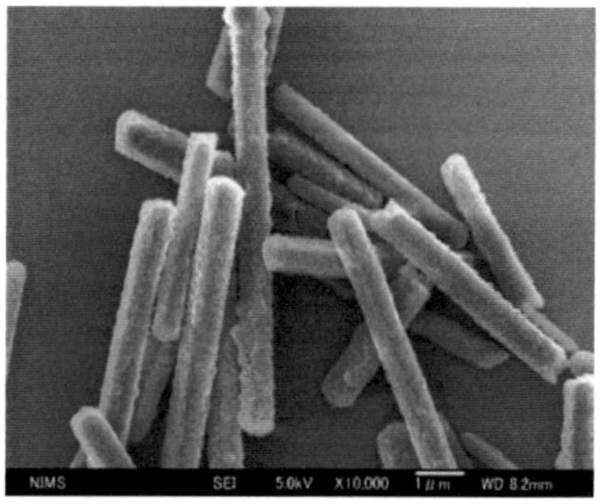

Figure 2. SEM image of C_{60}NWs.

3.1.2. Dispersion of C_{60}NWs in culture
Before exposuring cells to C_{60}NWs, the dispersion of C_{60}NWs in culture was examined visually by the inverted optical phase-contrast microscope for the same period. The suspensions of C_{60}NWs were poured into the cell-free medium by the same method as the exposure examination. The C_{60}NWs were well and stably dispersed onto the dishes of culture medium during observation (Fig.5).

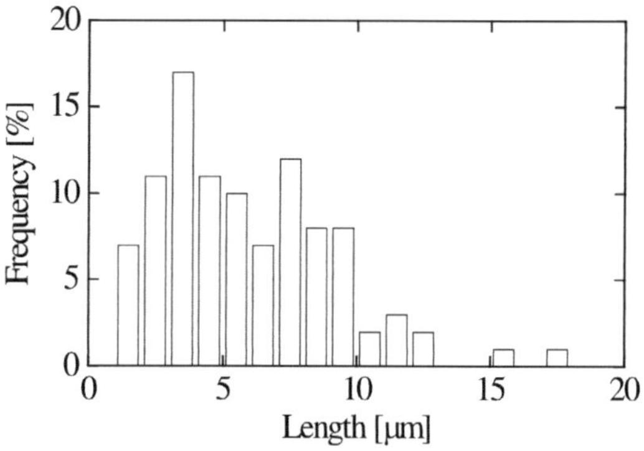

Figure 3. Histogram of C_{60}NWs' length.

3.2. Effect of PMA concentration and induction period on differentiation of THP-1 cells into macrophage-like cells

Optimal PMA concentration and induction period for the differentiation of THP-1 cells into macrophage-like cells were estimated under the condition of this study. In the case of 10 nM and 100 nM, cells adhered to the dish surface gradually (decrease of suspended cells) and about 90% of seeded cells adhered after 24 h of the PMA treatment (Fig.6). By contrast, the suspended cells grew by a factor of 5 in comparison with the seeded cells during 72 h in control culture (without PMA treatment). At the concentration of 1 nM, a few cells adhered the dish surface (data not shown), but the number of suspended cells increased about by a factor of 2 during 72 h. We observed that some of the cells showed elongation and pseudopodia formation after a few hours of the treatment by 10 nM and 100 nM (data not shown) of PMA, and most of the cells were observed to change their morphology by the PMA treatment of 24 h with the inverted optical phase-contrast microscope (Fig.7).

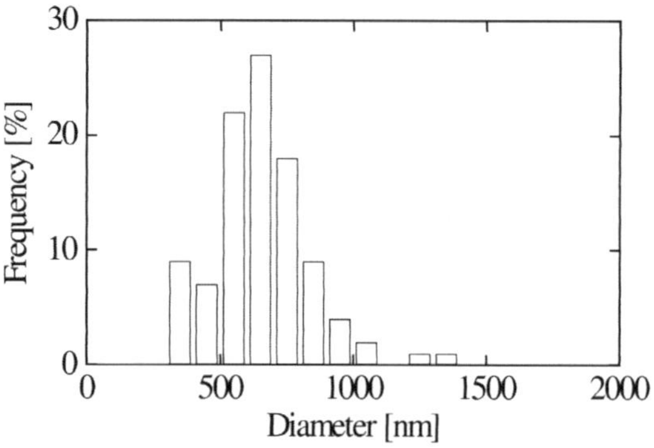

Figure 4. Histogram of C_{60}NWs' diameter.

Because no significant difference was observed in the cellular morphology and the number of adherent cells between 10 nM and 100 nM of PMA treatment, the cells differentiated by 10 nM PMA treatment for 24 h were used for the exposure experiment of C_{60}NWs.

3.3. Exposure experiments of C_{60}NWs

3.3.1. *Phagocytosis of C_{60}NWs*

Macrophages recognize, internalise and digest foreign materials. The uptake of them depends on their size and surface properties [5]. C_{60} is also phagocytized by macrophages [6] and the uptake rate of C_{60} is lower than that of graphite particles [7]. We observed that the macrophage-like cells gradually internalised C_{60}NWs after the exposure to C_{60}NWs with an inverted optical phase-contrast microscope (Fig.8). We will three-dimensionally locate the position of C_{60}NWs among the cells and study the phagocytosis of C_{60}NWs in more detail.

3.3.2. *Effect of C_{60}NWs' concentration and exposure period on the cell morphology.*

No alteration of cellular morphology was observed in the macrophage-like cells compared to the control group without exposure to C_{60}NWs after the exposure of 48 h for any concentration of C_{60}NWs in the observations by the inverted optical phase-contrast microscope (Fig.9). However, we are going to investigate the endpoints such as cell viability, cytokines, LDH and active oxygen generation in the next follow-up study using positive controls.

We cannot explain the fiber paradigm that a hazardous fibre is thinner than 3 μm in diameter, longer than 20 μm and biopersistent in the lungs [1] in the present paper dealing with only short (< 20 μm) C_{60}NWs. But some previous studies have reported that C_{60} (aggregate size was not described or larger than 1 μm) were nontoxic against mammalian cells [7, 8, 9] and dissolved inside lipid droplets in liver rats [10]. Owing to the weak van der Waals bonding forces acting between C_{60} molecules, C_{60}NWs may decompose into individual C_{60} molecules in living organisms. And on the basis of this assumption, C_{60}NWs may exhibit a biological effect like C_{60}. But these discussions are unclear now. We will carry out further researches on the digestion and dilution of C_{60}NWs using short and long C_{60}NWs.

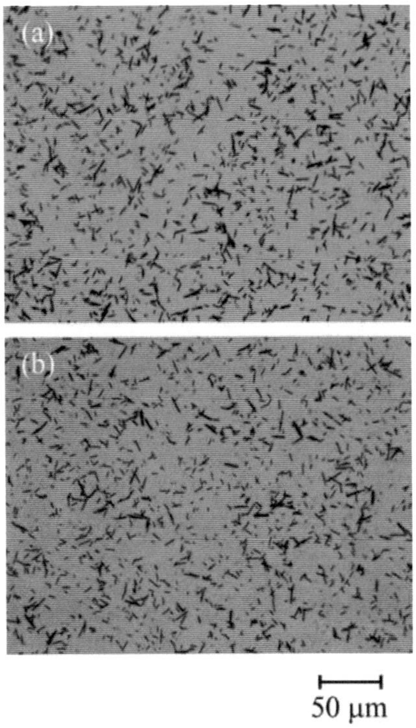

Figure 5. C_{60}NWs' dispersion in culture at a concentration of 10 μg/mL. (a) 1h and (b) 48 h after pouring the suspension of C_{60}NWs into the cell-free medium.

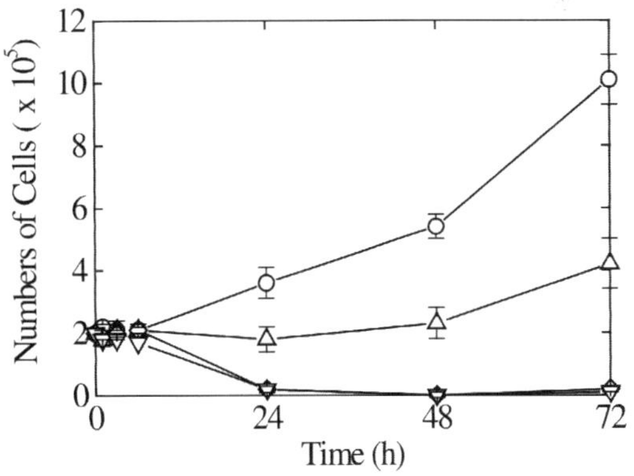

Figure 6. Suspended living cells after the PMA treatment. Circle, control (untreated); triangle, 1 nM PMA; diamond, 10 nM PMA; inverted triangle, 100 nM PMA. The results are expressed as the mean values for triplicate cultures with standard deviations.

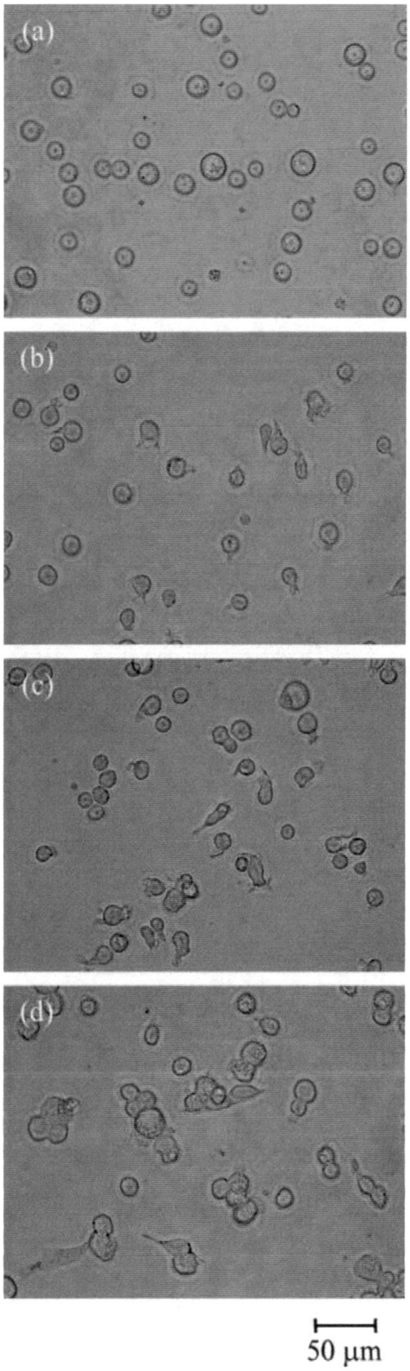

Figure 7. Morphological changes of THP-1 cells after (a) 1 h, (b) 3 h, (c) 6 h and (d) 24 h of 10 nM PMA treatment.

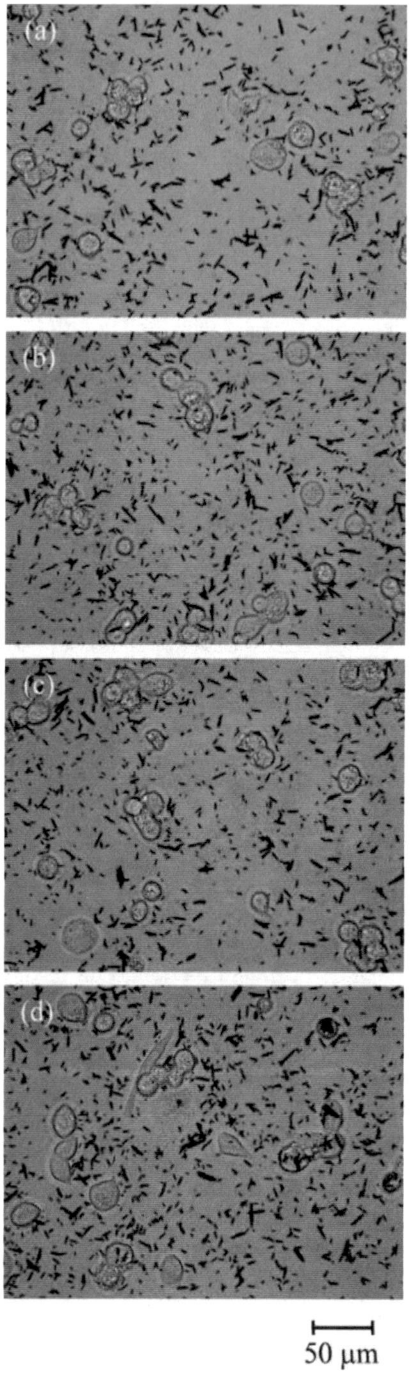

Figure 8. Macrophage-like cells exposed to C_{60}NWs for (a) 1 h, (b) 3 h, (c) 6 h and (d) 24 h at a concentration of 10 μg/mL.

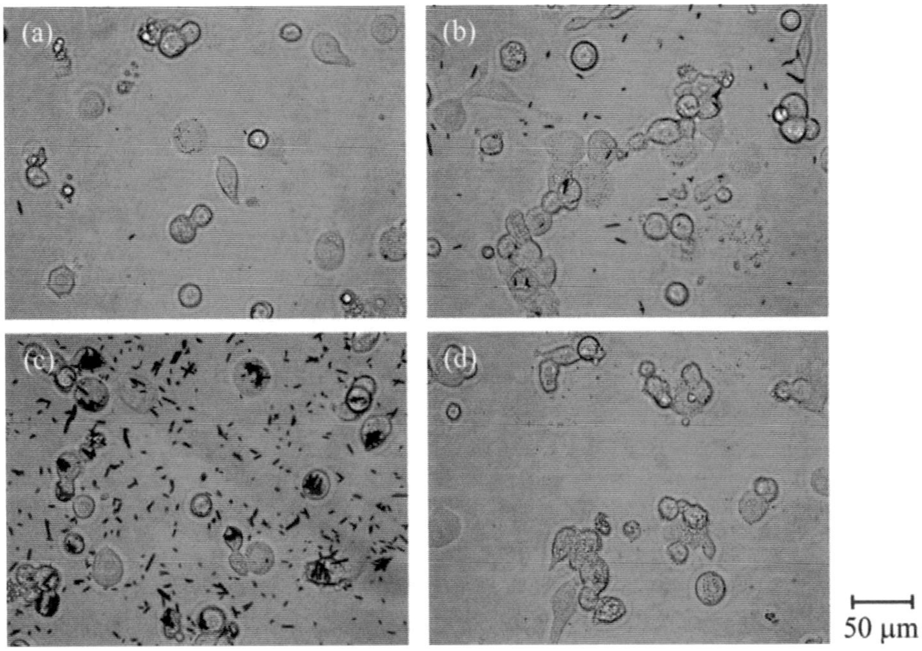

Figure 9. Macrophage-like cells exposed to C_{60}NWs for 48 h at concentrations of (a) 0.1 μg/mL, (b) 1 μg/mL and (c) 10 μg/mL. (d) Control culture cultivated without C_{60}NWs.

4. Conclusions

C_{60}NWs with an average length of about 6.0 μm and an average diameter of about 660 nm were well and stably dispersed onto the dishes of culture medium. Macrophage-like cells internalised the C_{60}NWs gradually, but no alteration of cellular morphology was observed in the macrophage-like cells for any concentration of C_{60}NWs (0.1, 1 and 10 μg/mL) compared to the control group without the exposure to C_{60}NWs for 48 h. We will complete this research for the biological impacts of C_{60}NWs in the next follow-up study using asbestos as a positive control and different sizes of C_{60}NWs and CNTs.

Acknowledgment
Part of this work was supported by NIMS Center for Nanotechnology Network.

References

[1] Poland C A, Duffin R, Kinloch I, Maynard A, Wallace W A H, Seaton A, Stone V, Brown S, Macnee W and Donaldson K 2008 *Nature Nanotechnol.* **3** 423
[2] Miyazawa K, Kuwasaki Y, Obayashi A and Kuwabara M 2002 *J.Mater.Res.* **17** 83
[3] Tsuchiya S, Yamabe M, Yamaguchi Y, Kobayashi Y, Konno T and Tada K 1980 *Int. J. Cancer* **26** 171
[4] Tsuchiya S, Kobayashi Y, Goto Y, Okumura H, Nakae S, Konno T and Tada K 1982 *Cancer Research* **42** 1530
[5] Tabata Y and Ikada Y 1988 *Biomaterials* **9** 356
[6] Porter A E, Muller K, Skepper J, Midgley P and Welland M 2006 *Acta Biomaterialia* **2** 409
[7] Fiorito S, Serafino A, Andreola F and Bernier P 2006 *Carbon* **44** 1100
[8] Moussa F, Chretien P, Dubois P, Chuniaud L, Dessante M, Trivin F, Sizaret P Y, Agafonov V, Ceolin R, Szwarc H, Greugny V, Fabre C and Rassat A 1995 *Fullerene Sci.Technol.* **3** 333
[9] Baierl T, Drosselmeyer E, Seidel A and Hippeli S 1996 *Exp. Toxic. Pathol.* **48** 508
[10] Gharbi N, Pressac M, Hadchouel M, Szwarc H, Wilson S R and Moussa F 2005 *Nano Lett.* **5** 2578

Formation and structures of C_{60} nanowhiskers with high strength and large elasticity

Mami Watanabe[1], Takahiro Yamamoto[1], Kenichi Kojima[1], Makoto Tanimura[2] and Masaru Tachibana[1*]

[1]International Graduate School of Arts and Sciences, Yokohama City University, 22-2 Seto, Kanazawa-ku, Yokohama 236-0027, Japan
[2]Research Department, *NISSAN ARC, LTD.*, Natsushima-cho, Yokosuka 237-0061, Japan

E-mail address: tachiban@yokohama-cu.ac.jp

Abstract. Formation and structure of toughened C_{60} nanowhiskers (C_{60}NWs) were investigated. C_{60}NWs are modified when pristine ones grown by a liquid-liquid interfacial precipitation method are kept in the growth solution for more than 3 years. The modified ones have a hexagonal solvated structure not only in the solution but also in air or a vacuum. The modified ones exhibit high strength and large elastic deformation in air, quite differing from the brittle in pristine ones in air.

1. Introduction

A new type of C_{60} needle-like single crystals called C_{60} nanowhiskers (C_{60}NWs) has been grown by a liquid-liquid interfacial precipitation (LLIP) method, where they nucleate at the interface between a toluene solution of C_{60} and isopropyl alcohol [1]. Typical C_{60}NWs have less than 500 nm in diameter and more than 100 μm in length. Especially, C_{60}NWs exceeding 1 mm in length can be easily obtained [2]. Such unique shapes of C_{60}NWs can provide us with various applications such as nano-scale electronic devices and field emitters. Some mechanical [3] and electrical properties [4] of C_{60}NWs have been already reported.

Pristine C_{60}NWs in the growth solution have a hexagonal solvated structure and their growth direction corresponds to a [0001] direction [5]. C_{60} molecules in the nanowhiskers are condensed with van der Waals force [2]. The hexagonal solvated structure is easily changed by drying, i.e. the evaporation of solvent molecules in air or a vacuum [5]. Consequently, the dried C_{60}NWs have a face-centered cubic (fcc) structure where the long axis corresponds to a [110] direction. The perfection in dried C_{60}NWs is low compared with pristine ones [6]. In addition, the dried ones are brittle and relatively soft. The mechanical weakness in C_{60}NWs in air or a vacuum is a severe problem for the practical applications. On the other hand, it seems that pristine C_{60}NWs in the growth solution have large elastic deformation and are relatively tough [6]. For the practical applications, it is desired that such mechanical toughness in pristine C_{60}NWs are kept even in air or a vacuum.

Recently we found that C_{60}NWs kept in the growth solution for more than 3 years show higher strength and larger elastic deformation in air. The production of such toughened C_{60}NWs might lead to

* To whom any correspondence should be addressed.

a breakthrough for the practical applications. In this paper, we report the formation and structure of toughened C_{60}NWs.

2. Experimental procedure

C_{60}NWs were grown by the LLIP method as reported previously [7]. C_{60}NWs grown for more than one month exceed 1 mm in length but are still below 500 nm in diameter. C_{60}NWs grown for about 1 week to several months were defined as "pristine" specimens. Moreover, C_{60}NWs kept in the growth solution for more than 3 years was called as "3-years" ones. These two kinds of specimens were investigated in this work. The LLIP method included some C_{60} bulk crystals as impurities. They predominantly contribute to X-ray diffraction (XRD) and Raman spectra. To obtain intrinsic XRD and Raman spectra of C_{60}NWs, specimens were purified by several filtrations. The purified ones were used in the measurements.

The surface of C_{60}NWs was observed by scanning electron microscopy (SEM) (S-4700, Hitachi). The structure and intermolecular interaction of C_{60}NWs were examined by XRD (RINT X-ray Diffractometer, Rigaku) and Raman spectroscopy (FRA 106/S, Bruker). To clarify their features in the growth solution and air, time evolution of the XRD and Raman spectra of C_{60}NWs during the drying was measured. In the XRD measurements, the specimens were mounted on a glass sample holder immediately after the sampling from the growth solution. XRD patterns were taken with a 0.1542 nm of Cu Kα. In the Raman measurements, the specimens were mounted on a mirror immediately after the sampling from the growth solution. The Raman spectra were taken with a 1064 nm excitation from a YAG laser in order to avoid the photo-polymerization of C_{60} by the excitation.

The mechanical properties of 3-years ones in air were investigated by a bending method. To observe the bending behavior, the specimens were mounted on the glass plate with droplets, from the growth solution. The bending deformation for specimens was carried out by applying a wind pressure with a fan as reported previously [6]. In order to apply larger stress, the tip of specimen was manually

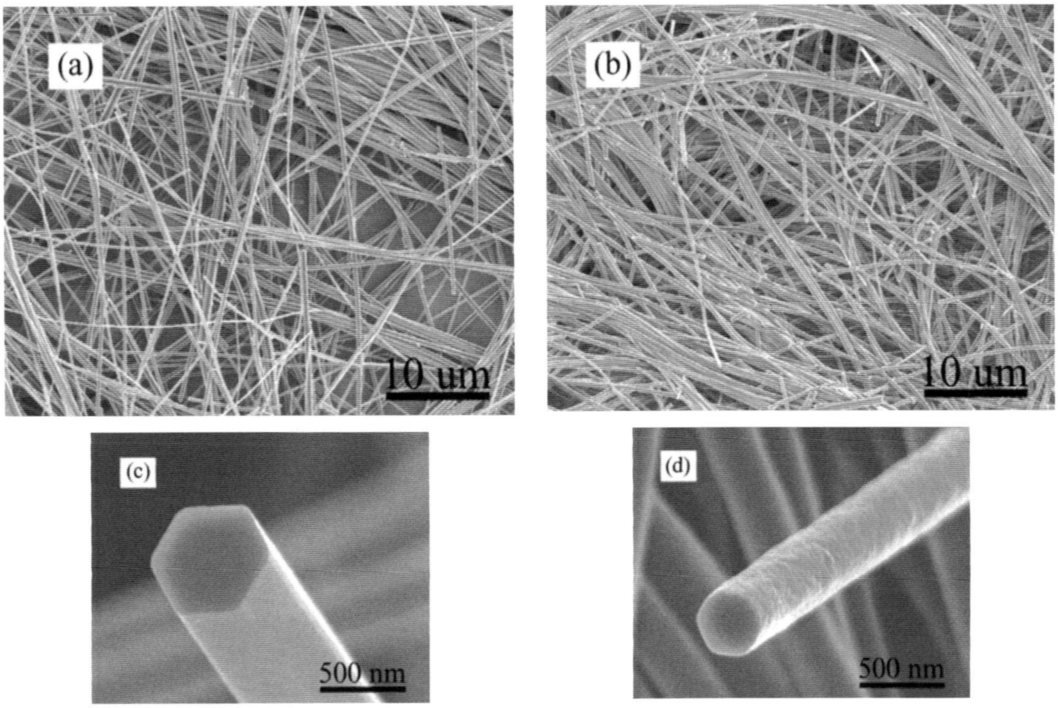

Figure 1. SEM images of (a) pristine C_{60}NWs, and (b) 3-years ones that were formed when pristine ones were kept in the growth solution for more than 3 years. (c) and (d) show expanded figures for (a) and (b), respectively.

pushed by using a needle. However, the stress applied to the specimen could not be measured in this experiment. The observation of the bending behavior was carried out in-situ by using an optical microscope (BX51, Olympus) with a charge coupled device (CCD).

3. Results and discussion

Figure 1(a) and 1(b) show SEM images of pristine C_{60}NWs, and 3-years ones that were formed when pristine ones were kept in the growth solution for more than three years, respectively. As shown in figure 1(a) and 1(b), straight and curved C_{60}NWs are clearly observed for both specimens. Some bundles are also observed. In addition, as seen in the expanded figures of figure 1(c) and 1(d), the shape of cross section of each C_{60}NW is found to be hexagonal-like one for both specimens. However, the surface in 3-years ones in figure 1(d) is much rough compared with that in pristine ones in figure 1(c). According to the preliminary gas chromatograph mass spectrometric analysis, oxides originated from toluene and some other products were observed only for 3-years ones. Therefore, the surface roughness might be due to some chemical reactions such as oxidation with soluble oxygen. In addition, it is expected that amorphous layers are thickly formed in the surface of 3-years ones.

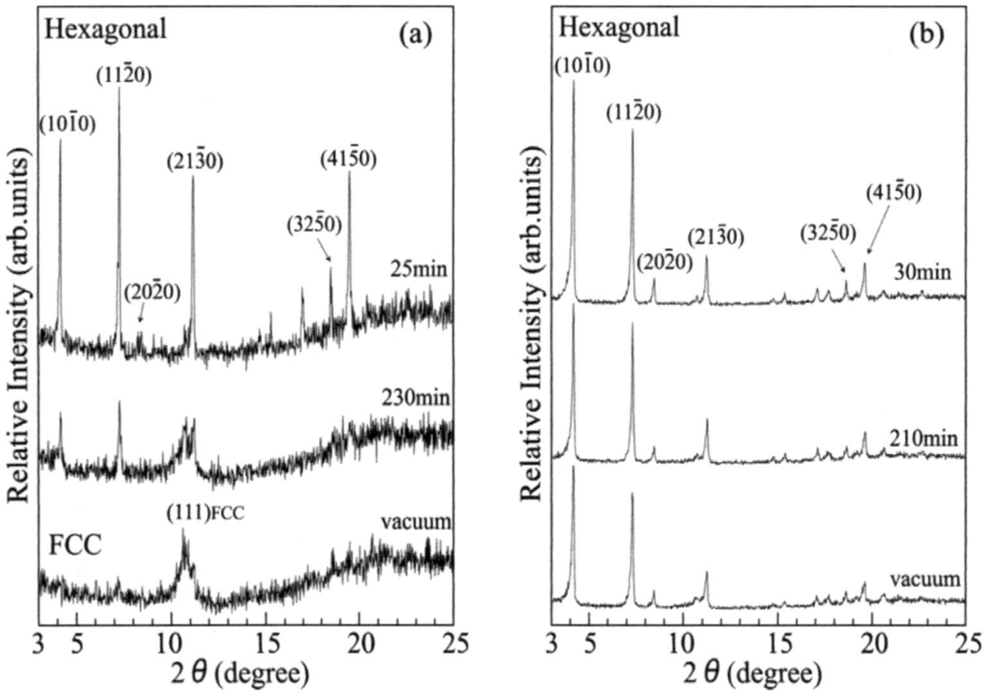

Figure 2. Time evolution of XRD pattern during the drying in air for (a) pristine C_{60}NWs and (b) 3-years C_{60}NWs that were formed when pristine ones were kept in the growth solution for more than 3 years. (a) and (b) also include XRD patterns after drying in a vacuum. The XRD patterns were obtained with 0.1542 nm of Cu Kα.

For the pristine and 3-years ones, time evolution of XRD pattern during the drying was measured as shown in figure 2(a) and 2(b), respectively. As shown in figure 2(a), for pristine C_{60}NWs, XRD pattern taken for 25 min after the sampling from the glass bottles showed sharp peaks. The peaks were indexed by a hexagonal system with lattice constants of a=2.43 nm and c=1.01 nm (a/c=2.41). The hexagonal peaks are also in good agreement with those of C_{60} solvated structure as reported previously [5].The hexagonal solvated structure for 25 min can correspond to pristine ones in the growth solution.

As shown in figure 2(a), these sharp peaks were observed even after air-drying for 230 min. In the XRD pattern for 230 min, the broad peaks were also appeared at similar positions to those for the intrinsic C_{60} crystals, and indexed by a fcc system with a lattice constant of a=1.44 nm. Note that the peak broadening is due to the lowering of crystal quality for dried ones. In addition, almost all hexagonal peaks disappeared after drying in a vacuum so that only fcc broad peaks were observed as shown in figure 2(a). This indicates that the hexagonal solvated structure in pristine C_{60}NWs is easily changed into the fcc, by drying. Such a change in the structure is due to the evaporation of solvent molecules in air or a vacuum. Similar change has been also observed in previous reports [5,6].

Figure 2(b) shows the time evolution of XRD pattern of 3-years ones during the drying. The XRD pattern taken for 30 min in air exhibits the peaks corresponding to the hexagonal solvated structure in pristine ones for 25 min in figure 2(a). The peaks were indexed by a hexagonal system with lattice constants of a=2.41 nm and c=1.01 nm (a/c=2.39). The lattice constant of a-axis is slight small compared with that of pristine ones as mentioned above. It should be noticed that no significant change in the XRD pattern was observed even after air-drying for 210 min as seen in figure 2(b). The hexagonal XRD pattern was also kept even after drying in a vacuum of 10^{-5} Pa for 4 hours. Therefore, the hexagonal solvated structure for 3-years ones is kept even in air or a vacuum. In addition, no peak broadening for 3-years ones was observed with drying. This means that the high quality of 3-years ones is kept even after drying. Such structural feature for 3-years ones is quite different from that for pristine ones as mentioned above. This can be attributed to the difficulty of the evaporation of solvent molecules due to the thicker amorphous layer formed in the nanowhisker surface.

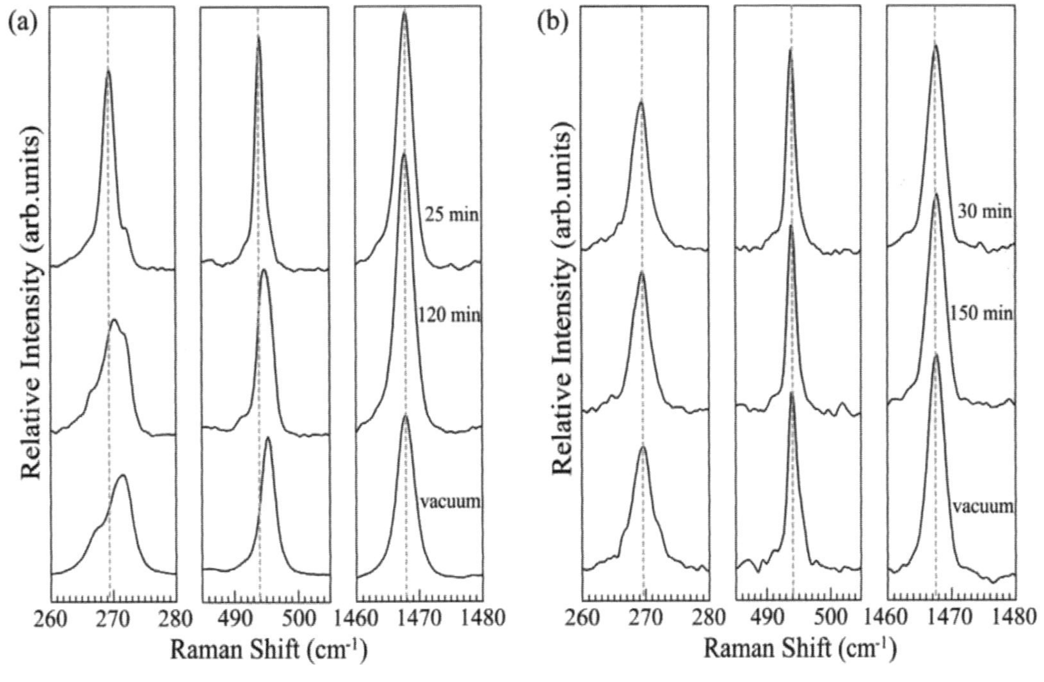

Figure 3. Time evolution of main Raman bands during in air for (a) pristine C_{60}NWs and (b) 3-years ones that were formed when pristine ones were kept in the growth solution for more than 3 years. (a) and (b) also include Raman bands after drying in a vacuum. The Raman bands at 270, 494, and 1468 cm^{-1} can be associated with $H_g(1)$ "squashing", $A_g(1)$ "breathing", $A_g(2)$ "pentagonal pinch" modes of C_{60} molecules, respectively. The Raman spectra were taken with 1064 nm excitation.

The time evolution of the Raman spectra for pristine C_{60}NWs and 3-years ones during the drying was also measured to examine the intermolecular interaction in the growth solution and air. Figure 3 shows the time evolution of main Raman bands at 270, 494, and 1468 cm^{-1}, which can be associated with $H_g(1)$ squashing, $A_g(1)$ breathing, $A_g(2)$ pentagonal pinch modes of C_{60} molecules, respectively. It is known that the peak profile of $H_g(1)$ for C_{60} molecules is quite different from that for crystalline C_{60}. Namely, the peak for C_{60} molecules can be fitted by one Lorentzian curve while that for the crystals is fitted by several curves. Such splitting in $H_g(1)$ mode for the crystals should be attributed to the crystal field due to the regular molecular arrangement in the crystals.

As discussed above, C_{60}NWs for 25 min in air correspond to pristine ones in the growth solution, and have the hexagonal solvated structure. However, for the corresponding Raman spectrum for 25 min as shown in figure 3(a), no clear splitting in $H_g(1)$ peak was observed. This means that pristine C_{60}NWs in the growth solution have a smaller crystal field, i.e. smaller intermolecular interaction although they have the regular molecular arrangement. The small crystal field would be due to the solvent molecules in the crystals. On the other hand, clear splitting in $H_g(1)$ peak was observed for fcc dried ones after air-drying for 120 min or after drying in a vacuum as shown in figure 3(a). This can be attributed to the increase of the crystal field of fcc lattice due to the evaporation of solvent molecules in air or a vacuum. Moreover, it was found that the peak frequency of $A_g(1)$ was slightly up-shifted with drying. This can be also explained by the increase of the intermolecular interaction due to the evaporation of solvent molecules.

Figure 3(b) shows the time evolution of the Raman spectrum of 3-years ones during the drying. The frequency and shape of Raman peaks of 3-years ones taken for 30 min in air are similar to those of pristine ones for 30 min in air in figure 3(a). This shows that no strong chemical bond of C_{60} with C_{60} or others such as solvents is formed even for 3-years ones, as pristine ones. Moreover, no change in the frequency and shape of all peaks for 3-years ones was observed even after air-drying for 150 min or after drying in a vacuum of 10^{-5} Pa for 4 hours. This means that 3-years ones in air and a vacuum can maintain the same intermolecular interaction as that in the growth solution. Such a feature for 3-years ones is quite different from that of pristine ones as mentioned above. These Raman features in figure 3 are also strongly correlated with those of XRD in figure 2.

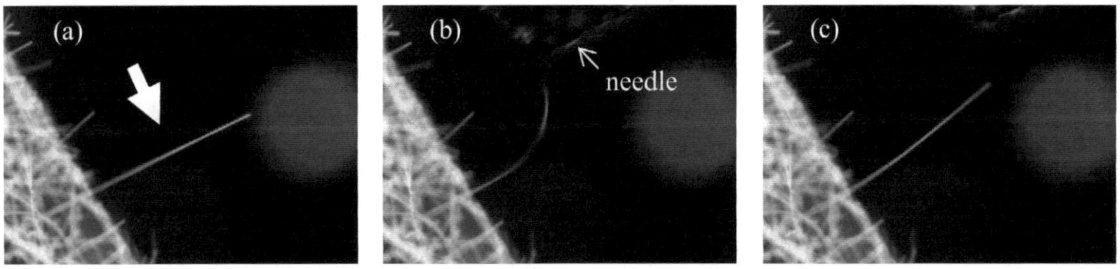

Figure 4. Successive pictures of the bending deformation in air for a 3-years C_{60}NW that was formed when pristine one was kept in the growth solution for more than 3 years. The pictures show (a) straight one before applying a stress, (b) bent one with a curvature of 80 μm by applying a stress, and (c) straight one recovered after freeing from a stress, respectively. Note that the stress was manually applied to the tip of specimen with a needle as seen in (b).

For 3-years ones, the bending experiments were carried out in air. Figure 4 shows the successive pictures of the bending behavior of a 3-years one which has 500 nm in diameter and more than 100 μm in length. The straight one before applying a stress is indicated by the arrow in figure 4(a). According to the previous report [6], a wind pressure with a fan led to the break of pristine C_{60}NWs in air. However, 3-years ones in air were not broken by the same wind pressure as the previous one. This means that 3-years ones have high strength even in air, at least compared with pristine ones as reported previously [6]. In order to apply larger stress, the tip of specimen was manually pushed by using a needle. Consequently, the bending deformation clearly occurred. Namely, the 3-years one was bent with a curvature of 80 μm, as shown in figure 4(b). The curvature is smaller than that for pristine ones in the solution [6]. Freeing from the stress, it almost recovered to original shape although a small plastic deformation occurs, as shown in figure 4(c). Such behavior of the bending shows that 3-years ones exhibit large elastic deformation even in air, as that in pristine ones in the growth solution [6]. This is because the hexagonal solvated structure for the 3-years ones is kept even in air or a vacuum. In addition, the toughness might be attributed to the thicker amorphous layer in the surface of 3-years ones.

4. Conclusion

C_{60}NWs are modified when pristine ones grown by a liquid-liquid interfacial precipitation method are kept in the growth solution for more than 3 years. The modified ones have the hexagonal solvated structure not only in the solution but also in air or a vacuum. The modified ones exhibit high strength and large elastic deformation even in air, quite differing from the brittle in pristine ones in air. The production of such toughened C_{60}NWs might lead to a breakthrough for the practical applications.

Acknowledgment

The authors thank Dr. K. Miyazawa of National Institute for Materials Science for the bending experiments. This work was partially supported by a Strategic Research Project (T20007) in Yokohama City University.

References

[1] Miyazawa K, Kuwasaki Y, Obayashi A and Kuwabara M 2002 *J. Mater. Res.* **17** 83-88
[2] Tachibana M, Kobayashi K, Uchida T, Kojima K, Tanimura K and Miyazawa K 2003 *Chem. Phys. Lett.* **374** 279-285
[3] Asaka K, Kato R, Miyazawa K and Kizuka T 2006 *Appl. Phys. Lett.* **89** 071912
[4] Ogawa K, Kato T, Ikegami A, Tsuji H, Aoki N, Ochiai Y and Bird J P 2006 *Appl. Phys. Lett* **88** 112109
[5] Minato J and Miyazawa K 2005 *Carbon* **43** 2837-2841
[6] Watanabe M, Miyazawa K, Kojima K and Tachibana M 2008 *IEEJ Trans. SM* **128** 321-324
[7] Kobayashi K, Tachibana M and Kojima K 2005 *J. Cryst. Growth* **274** 617-621

GC-MS analysis of the solvents contained in C_{60} nanowhiskers

M Watanabe[1], K Hotta[2], K Miyazawa[2] and M Tachibana[1]

[1]International Graduate School of Arts and Sciences, Yokohama City University, 22-2 Seto, Kanazawa-ku, Yokohama 236-0027, Japan
[2]Fullerene Engineering Group, Advanced Nano Materials Laboratory, National Institute for Materials Science, 1-1 Namiki, Tsukuba, Ibaraki 305-0044, Japan

E-mail address: tachiban@yokohama-cu.ac.jp

Abstract. The solvent molecules contained in C_{60}NWs affect not only the shape but also the crystal structure of C_{60}NWs. The solvents contained in C_{60}NWs were found to be toluene and IPA by the GC-MS analysis. The most abundant solvent was toluene and the quantity of residual toluene in the as-prepared C_{60}NWs was about 7 mass%. The quantity of impurity solvents is reported for the as-prepared, air-dried, vacuum-dried and heat-treated C_{60}NWs.

1. Introduction

Solution-grown C_{60} crystals have various structures and shapes that depend on organic solvent, growth temperature and so on. Recently, a new type of C_{60} needle-like crystals called C_{60} nanowhiskers (C_{60}NWs) were obtained in a solution by the liquid-liquid interfacial precipitation (LLIP) method, where the C_{60} nanowhiskers nucleate at the interface between a toluene solution of C_{60} and isopropyl alcohol [1]. Typical C_{60}NWs have diameters less than 500 nm and lengths greater than a few micrometers. The high aspect ratio of C_{60}NWs with the small diameter is very unique, compared with those of any needle-like C_{60} crystals reported previously.

Same as C_{60} crystals, the shape of C_{60}NWs varies depending on the growth solution and growth condition [2-5]. For example, fullerene nanotubes, nanoporous fullerene nanowhiskers and fullerene nanosheets are synthesized by use of pyridine, benzene and CCl_4, respectively. The C_{60}NWs in solution have a hexagonal solvated structure and their growth direction is parallel to the [0001] direction. The hexagonal solvated structure rapidly changes into the face-centered cubic (fcc) structure by drying through the evaporation of solvent molecules in air or in vacuum [6]. The solvent molecules included in C_{60}NWs affect not only the shape but also crystal structure. However, there are really few studies about the solvents contained in C_{60}NWs [7, 8]. In this report, the solvent molecules contained in C_{60}NWs are examined by use of GC–mass spectrometry (GC-MS) for the as-prepared specimens, the specimens dried in vacuum and the heat-treated specimens.

2. Experimental

2.1. Samples

C_{60}NWs were grown by the LLIP method as reported previously [1]. A toluene (99.5% Wako Ltd.) solution saturated with C_{60} (99.95 % MTR Ltd.) was prepared and poured into a glass bottle. Next, isopropyl alcohol (IPA, 99.8% Wako Ltd.) was gently added to the solution to form a liquid-liquid interface. After the addition of IPA, the solution was manually mixed for a few times and was kept in an incubator (SANYO MIR-153) at 20 °C for 24 h.

2.2. Drying and heat treatment of samples

The C_{60}NWs grown in solutions were dispersed in IPA and collected by the centrifugal method. The C_{60}NWs were treated by three methods before the measurement by GC-MS. The first method is air drying, the second method is vacuum drying in a vacuum furnace and the third method is the heat treatment in Ar gas. The samples were dried in vacuum of a range of 10^{-2} and 10^{-4} Pa for 5 ~ 720 min. In the heat treatment in Ar, the air in the vacuum furnace was substituted by pure Ar (99.999%), and the C_{60}NWs were heated at 50 °C to 300 °C for 30 min. The C_{60}NWs were put in disposable containers (Eco-cap, Frontier Laboratories Ltd.) and heat-treated in Ar.

2.3. Measurement conditions by GC-MS

A thermal desorption (TD) method was used to measure the solvents contained in C_{60}NWs. Since C_{60} sublimates at about 300 °C in vacuum [9], the heating temperature was set at 350 °C. It was expected that the solvent molecules contained in the matrix of C_{60}NWs were instantaneously detached at 350 °C, where the TD system with Double-Shot Pyrolyzer (Frontier Laboratories Ltd. PY-2020iD) was used. The gases from C_{60}NWs were analyzed by GC-MS system (SHIMADZU GCMS-2010) using a split rate of 1:100. The gas chromatography (GC) was performed at an injection temperature of 330 °C, using a helium carrier gas pressure and a column composed of 5% diphenyl and 95% dimethylpolysiloxane (Ultra ALLOY+, Frontier Laboratories Ltd.). The oven was heated from 40 °C to 300 °C at a heating rate of 30 °C min^{-1}. The mass analysis was done by using the EI+ ionization mode at an interface temperature of 330 °C.

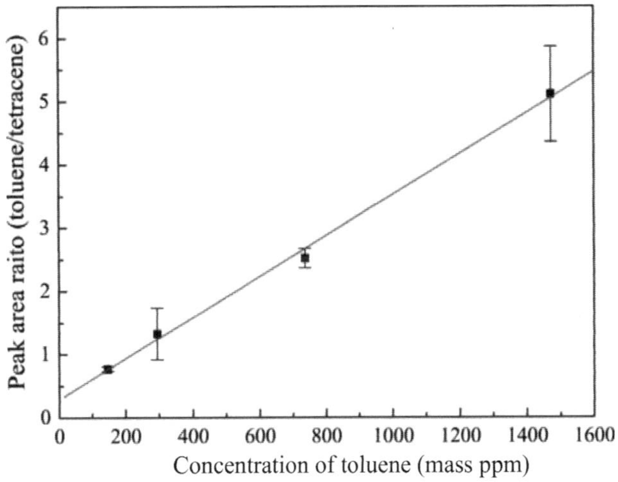

Figure 1. The calibration curve for the quantitative analysis of toluene. The horizontal axis expresses the concentration of toluene (mass ppm). The vertical axis expresses the peak area ratio of toluene versus tetracene standard.

About 300 ~ 550 μg of C_{60}NWs were collected by centrifugation and used in each measurement by GC-MS. Before the GC-MS analysis, the sample cases containing the C_{60}NWs other than air drying were dried in vacuum for 5 min at room temperature in order to remove the solvents adsorbed on the surface of C_{60}NWs. Tetracene was used as a standard material. 0.0025 g of tetracene was dissolved in 50 ml of hexane. 100 μL of tetracene-hexane solution was put into a sample case and heated at 50 °C on a heater in order to evaporate only hexane and about 5 μg of tetracene remained in the sample case. The C_{60}NWs dried in vacuum for 5 min were put into the same sample case containing 5 μg tetracene and subjected to the GC-MS analysis. The other samples were also similarly treated and subjected to the GC-MS analysis. A calibration curve for the quantitative analysis of toluene was made to measure the amount of residual toluene contained in C_{60}NWs as shown in Figure 1. The horizontal axis expresses the concentration of toluene (mass ppm). The vertical axis expresses the peak arca ratio of toluene versus the tetracene standard. 1, 2, 5 and 10 mg of toluene were diluted with 10 ml of hexane. The calibration curve was made by using 2 μL of the hexane solutions of toluene with the concentrations of 1, 2, 5 and 10 μg μL^{-1}, and 5 μg of tetracene for GC-MS at the same time. The amount of residual toluene contained in C_{60}NWs was measured with the calibration curve.

3. Results and discussion

Figure 2 shows a GC-MS analysis of C_{60}NWs after drying in vacuum for 5 minutes. In figure 2(a), there are four peaks labeled at 0.925, 1.067, 2.33 and 10.09 min. In Figure 2(a), the peak of 0.925 min is consistent with the ions originated from CO_2, N_2 and water. Hence, the surface of C_{60}NWs must have adsorbed water, CO_2 and N_2. But the peak may contain the air introduced into the GC-MS system at the time of sample injection. All of the measured CO_2, N_2 and water may not have been from the surface of C_{60}NWs. Figure 2(b) shows the mass spectra for the peak labeled at 2.33 min with the major ions at 91 and 92 Da. The molecular weight of toluene is 92, and the peak of 91 Da shows the fragments formed by the elimination of hydrogen atoms from toluene molecules in the electron ionization chamber. Figure 2(c) shows the mass spectra for the peak labeled at 1.067 min with the major ions at 45, 59 and 43 Da. The peak of 45 Da shows the fragments of isopropyl alcohol molecules formed by the elimination of $-CH_3$ groups and the peak of 43 Da shows the fragments formed by the elimination of –OH groups. The peak labeled at 10.09 min is due to the tetracene standard. The amount of residual toluene is more abundant than IPA. This is consistent with the fact that C_{60} is hardly dissolved in polar solvents like alcohol. Next, the most abundant toluene contained in the matrix of C_{60}NWs is quantitatively analyzed by use of the peak area ratio of toluene versus tetracene.

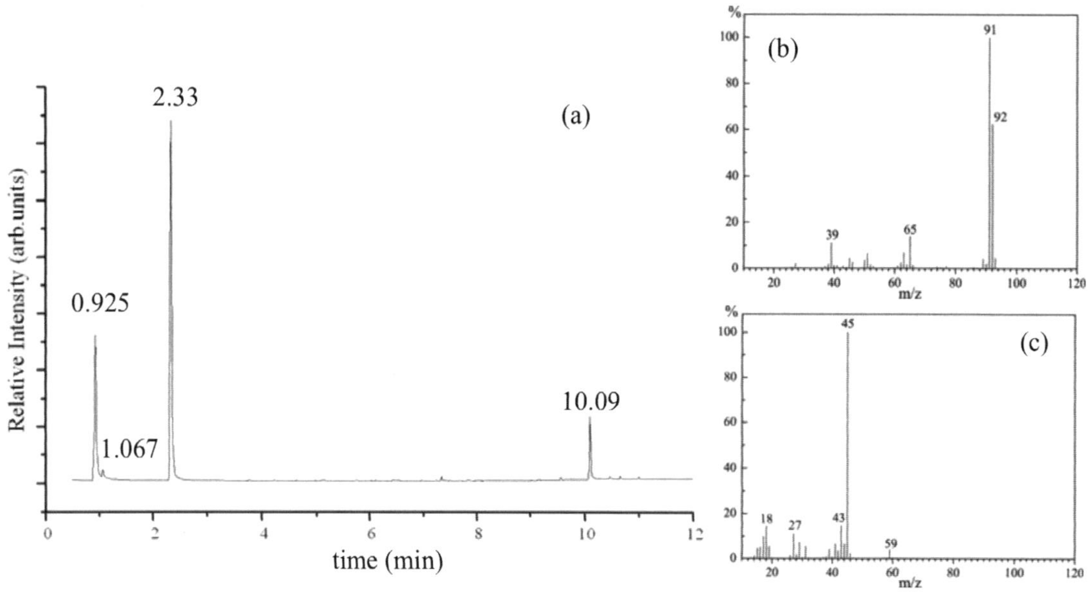

Figure 2. GC-MS analyses of C_{60}NWs. (a) GC spectrum of C_{60}NWs after drying in vacuum for 5 min, (b) the mass spectra corresponding to the peak of 2.33 min in (a) and (c) the mass spectra corresponding to the peak of 1.067 min in (a).

The amount of residual toluene contained in the vacuum-dried C_{60}NWs is roughly in proportion to the sample weight of C_{60}NWs. Figure 3 shows the amount of toluene measured by using the analytical curve of Figure 1 as a function of the weight of C_{60}NWs. Figure 3 shows that the amount of residual toluene in each sample of C_{60}NWs increases linearly with increasing the weight of C_{60}NWs.

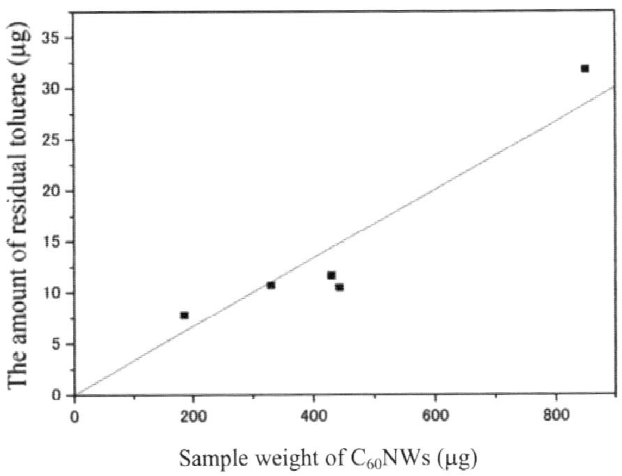

Figure 3. The amount of residual toluene contained in the C_{60}NWs measured by use of the calibration curve of Figure 1 as a function of the sample weight of C_{60}NWs dried in vacuum for 5 min at room temperature.

Next, the change of the toluene content in the C_{60}NWs (mass %) subjected to the air drying and that of the C_{60}NWs dried in vacuum at room temperature are shown in Figure 4. The C_{60}NWs were dried for 30, 60, 120 and 240 min in air, respectively. The toluene content for the as-prepared C_{60}NWs is also shown. The content of residual toluene in the C_{60}NWs shows similar values for the drying time of 30 to 240 min. The as-prepared C_{60}NWs show about 7 mass % of toluene content. It has been reported that hexagonal solvated structures of C_{60}NWs rapidly change into fcc structures upon drying through the evaporation of solvent molecules [6], corresponding to the present decrease of toluene in the C_{60}NWs. Figure 4 also shows the data of C_{60}NWs dried in vacuum at room temperature for 5, 120 and 720 min, respectively. The quantity of residual toluene decreased to about 1 % from 3.2 % in the C_{60}NWs that were dried in vacuum at room temperature for 120 min. But the residual toluene was not completely removed even for the elongated vacuum drying of 720 min. Hence, part of the toluene molecules contained in the matrix of C_{60}NWs must be tightly trapped by some special sites of C_{60} crystal lattices.

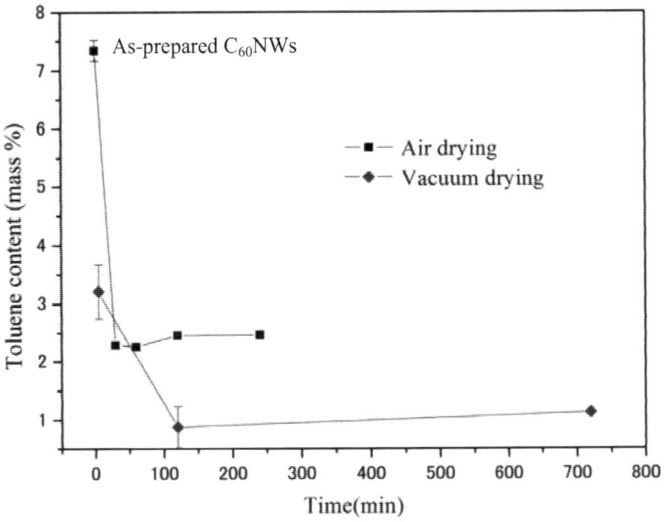

Figure 4. Change of the content of residual toluene in the dried C_{60}NWs as a function of drying time. Filled rectangles (■) show the residual toluene contained in the C_{60}NWs after air drying at room temperature for 30, 60, 120 and 240 min, respectively, and that of the as-prepared C_{60}NWs. Filled diamonds (◆) show the residual toluene contained in the C_{60}NWs dried in vacuum at room temperature for 5, 120 and 720 min.

Figure 5 shows the residual toluene content of the C_{60}NWs heated at 50, 100, 200 and 300 °C for 30 min in Ar gas and that of the C_{60}NWs dried in vacuum for 5 min at room temperature. The toluene content in the C_{60}NWs decreased to about 0.2 % from the initial content of 3.2 % rapidly by the heat treatment at 100 °C. Since the C_{60}NWs heat-treated at high temperatures turn to porous structures [12], the contained solvent molecules are expected to escape easily through the formed pores. However, the residual toluene content shows the similar values in the C_{60}NWs heat-treated at 100, 200 and 300 °C. This result also suggests that part of the contained toluene molecules are tightly bound to some special sites of C_{60} lattices and cannot be easily removed by the heat treatment.

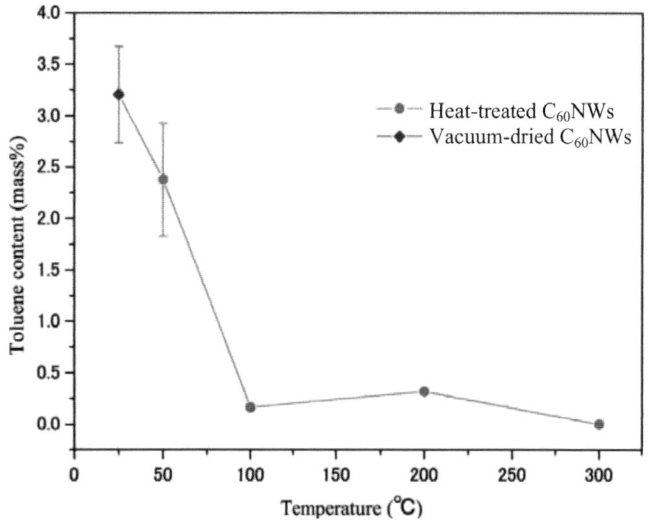

Figure 5. Change of the residual toluene content in the heat-treated or vacuum-dried C_{60}NWs expressed as a function of heating temperature. The C_{60}NWs were heated in Ar gas at each temperature for 30 min.

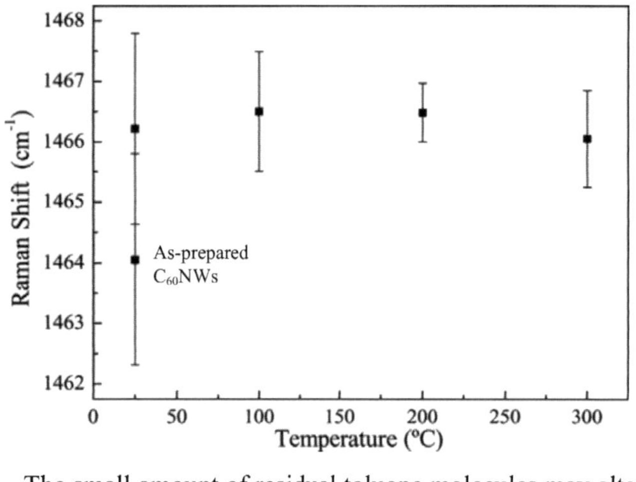

Figure 6. Raman shifts of $A_g(2)$ mode in the as-prepared C_{60}NWs, the C_{60}NWs dried in vacuum for 5 min at room temperature and the C_{60}NWs heated in Ar gas at 100 - 300 °C for 30 min.

The small amount of residual toluene molecules may alter the mechanical and electrical properties of C_{60}NWs from those of pure C_{60} crystals. In the Raman spectroscopy of C_{60}, it is known that the position of $A_g(2)$ peak is very sensitive to the polymerization of C_{60} [10, 11]. In the C_{60} nanotubes prepared by use of a C_{60}-saturated pyridine solution and IPA, a remarkable difference in the peak position of $A_g(2)$ mode was observed between the as-prepared solvated C_{60} nanotubes and the C_{60} nanotubes heat-treated at 100 to 500 °C [12]. The position of $A_g(2)$ peak was also measured in the present experiment for the as-prepared and heat-treated C_{60}NWs as shown in Figure 6. The average value of the Raman shifts of the C_{60}NWs heat-treated at 100 to 300 °C and that of the C_{60}NWs dried in vacuum at room temperature is 1466.2 $\pm$ 0.25 cm^{-1}, while the as-prepared solvated C_{60}NWs show the Raman shift of 1464.1 $\pm$ 1.7 cm^{-1}. From the above results of GC-MS spectrometry, it is found that the difference in the Raman shift between the as-prepared C_{60}NWs and the heat-treated C_{60}NWs is caused by the difference in the amount of solvent molecules contained in the C_{60}NWs. The weak peak shift of $A_g(2)$ mode in the as-prepared C_{60}NWs from 1469 cm^{-1} of pristine C_{60} [11] must be owing to a slight distortion of C_{60} molecules from their spherical icosahedral symmetry that is caused by the interaction between the C_{60} molecules and the surrounding solvent molecules.

4. Conclusions

The present research can be summarized as follows.

(1) Toluene was most abundantly contained in the C_{60}NWs.

(2) The content of toluene was 7 % in the as-prepared C_{60}NWs and rapidly decreased upon the heat treatment in vacuum or drying in the air.

(3) The down shift of $A_g(2)$ peak observed in the as-prepared solvated C_{60}NWs is assumed to be caused by the weak interaction between C_{60} molecules and the residual toluene molecules that deforms the spherical icosahedral symmetry of C_{60}.

In order to understand the mechanism of structural stability of solvated C_{60}NWs, it is necessary to study the time evolution between the amount of residual IPA and toluene contained in the C_{60}NWs and the crystal structural change of the C_{60}NWs upon drying. We are going to investigate this theme further.

References

[1] Miyazawa K, Kuwasaki Y, Obayashi A and Kuwabara M 2002 *J. Mater. Res.* **17** 83
[2] Miyazawa K, Minato J, Yoshii T, Fujino M and Suga T 2005 *J.Mater.Res.* **20** 688
[3] Sathish M, Miyazawa K and Sasaki T 2007 *Chem. Mater.* **19** 2398
[4] Sathish M and Miyazawa K 2007 *J. Am. Chem. Soc.* 129 45 13816
[5] Cha S I, Miyazawa K and Kim J 2008 *Chem. Mater.* **20** 1667
[6] Minato J and Miyazawa K 2005 *Carbon* **43** 2837
[7] Geng J, Zhou W, Skelton P, Yue W, Kinloch I, Windle A and Johnson B 2008 *J. Am. Chem. Soc.***130** 2527
[8] Miyazawa K, Chikashi N, Masahisa F, Tadatomo S and Tetsuro Y 2004 *Trans.Mat. Res. Soc. Jpn.* **29** 1965
[9] Cox D. M, Behal S, Disko M, Gorun S M, Greaney M, Hsu C S, Kollin E B, Millar J and Robbins J 1991 *J. Am. Chem. Soc.* **113** 2940
[10] Martin M. C, Koller D, Rosenberg A, Kendziora C and Mihaly L 1995 *Phys. Rev.* B **51** 3210
[11] Rao A M, Zhou P, Wang K A, Hager G T, Holden J M, Wang Y, Lee W T, Bi X X, Eklund P C, Cornet D S, Duncan M A and Amster I J 1993 *Science* **259** 955
[12] Miyazawa K, Ringor C, Nakamura K, Tachibana M, Saito K and Kizuka T 2008 *IEEJ Trans. SM* **128** 317

Tree-shaped C_{60}/C_{60}-Ferrocene Crystals by Kinetically Controlled LLIP Process

S I Cha[1], D Y Lee[1], K Miyazawa[2] and T Wakahara[2]

[1]Optoelectric Research Group, Korea Electrotechnology Research Institute, 70 Boolmosangil, Changwon 641-120, Korea
[2]Fullerene Engineering Group, National Institute for Materials Science, 1-1 Namiki, Tsukuba 305-0044, Japan

E-mail : sicha@keri.re.kr

Abstract. Among the one-dimensional C_{60} crystals, vertically aligned microtubes on the porous alumina substrates provide several interesting features including their well developed hexagonal cross sections and hollow internal structures. These characteristic structures of vertically aligned C_{60} microtubes are caused by their well developed crystals structures and can be predicted by classical nucleation and growth theory. Therefore, the variation of its chemical composition by doping or alloying or variation of kinetic environments during the growth can causes variety of structural changes from vertically aligned hexagonal microtubes. The morphology of vertically aligned C_{60} microtubes is modified by controlling kinetic factors and including of lattice defects. The controlling kinetic factors by modified injection mode of 2-propanol gives various inner structure of vertically aligned C_{60} microtubes such as filled inner structure, double walled inner structure and porous inner structures. The addition of lattice defects by inclusion of ferrocene doped C_{60} gives tree-shaped C_{60} crystals with many branches consisting of bundles of thin needles.

1. Introduction

C_{60} is an interesting form of carbon to both scientists and engineers [1-3] and recently discovered one-dimensional C_{60} nanostructures should bring about new applications for fullerene materials [4-8]. Several methods have been used to precipitate one-dimensional C_{60} crystals from solution, including nanowhiskers, nanotubes, and nanorods [2]. Among them, the liquid-liquid interface precipitation (LLIP) process has been used to produce long C_{60} nanowhiskers or nanotubes without templates. Moreover, recently introduced vertically aligned C_{60} microtubes on the porous alumina substrate provide new potential application fields by kinetically controlled LLIP process [9].

Among those one-dimensional C_{60} crystals, vertically aligned microtubes on the porous alumina substrates provide several interesting features including their well developed hexagonal cross sections and hollow internal structures. These characteristic structures of vertically aligned C_{60} microtubes are caused by their well developed crystals structures and can be predicted by classical nucleation and growth theory. Therefore, the variation of its chemical composition by doping or alloying or variation of kinetic environments during the growth can causes variety of structural changes from vertically aligned hexagonal microtubes.

In this study, the effect of kinetic variation during the growth and the effect of doping of ferrocene derivative of C_{60} to pure C_{60} crystals on the morphology are investigated. The variations of kinetic environments during the growth of vertically aligned C_{60} microtubes were controlled by controlling the injection of 2-propanol used for precipitation of C_{60} crystals from its solution. In addition, the inserting of small amount of ferrocene doped C_{60} induces the morphological changes of C_{60} crystals hence provide 3-dimensional tree-shaped C_{60} crystals.

2. Experimental Procedures

2.1. Preparation of solutions

Saturated C_{60} solutions in toluene (dehydrated 99.5%, organic synthesis grade) was prepared by mixing an excess amount of C_{60} powder (99.95% sublimated, MTR Ltd.) with toluene, followed by ultrasonication for 30 min. After sonication, the solutions were filtered through 450 nm Teflon syringe filters (Puradisc^{TM}, Whatman Inc). The solution containing 6 mol. % of ferrocene containing C_{60} derivatives, as shown in Scheme 1, and C_{60} in toluene, called ferrocene doped C_{60} solution, was prepared with same route to saturated C_{60} solution.

2.2. Preparation of vertically aligned C_{60} and ferrocene doped C_{60} crystals

In order to fabricate vertically aligned C_{60} and crystals and ferrocene doped C_{60} crystals, a glass filter holder for filters with a diameter of 25 mm was used. A syringe pump was connected to the lower part of filter holder by Teflon tubes. Anodized alumina (AAO) filters (Anodisc 25, 200 nm pore size, Whatman Inc.) were loaded in the filter holder. The solution (2 mL) was mixed with a 2-propanol with controlled rate using syringe pump. The injection rate was 0.02ml/min for vertically aligned crystals. All procedures were conducted in an incubator at 5 ^{o}C under ambient light. The amount of injected precipitator was 6 mL for 2 mL of the C_{60} solution. The separated C_{60} crystals were dried in an incubator at 5 ^{o}C in air. SEM micrographs were obtained on a FE-SEM (Hitachi-4800, 10 kV), and HRTEM and STEM were performed on a JEOL (JEM-2100F, 160 kV).

3. Results and Discussions

3.1. Effect of kinetic environments on the morphology of vertically aligned C_{60} crystals

The vertically aligned C_{60} microtubes can be prepared on the porous alumina substrate as reported before.[3] In this result, on interesting feature is that the outer surface of each microtube is smooth, while the inner surfaces are serrated. The reason is not clear, but it gives possibility to control inner structures.

The inner structure of vertically aligned C_{60} microtubes can be controlled by changing the injection mode. When the 2-propanol was injected continuously, the vertically aligned C_{60} microtubes with smooth outer surfaces and serrated inner surfaces are formed. One interesting feature is that when the 2-propanol is injected continuously until the vertically aligned C_{60} microtubes are formed, which is actually 4 hours injection with rate of 0.02ml/min. to 2ml of saturated C_{60} solution in toluene, and

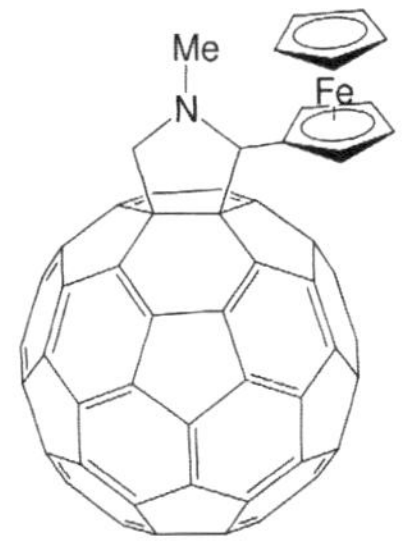

Scheme 1

Figure 1. SEM morphology of vertically aligned C_{60} microrods prepared by injection of 2-propanol for 4hours and incubated for 4 hrous to the C_{60} /toluene saturated solution

followed by holding 2-propanol injection and having incubation time around 4 hours, the inner space of microtubes are filled and they turns into microrods as shown in Fig. 1. It gives an interesting fact that the outer surfaces are formed first and then the inner structures are formed. Considering the serrated structures of inner space in vertically aligned C_{60} microtubes, the inner structures can be modified by changing injection mode.

When 2-propanol was injected with modified mode as shown in Fig. 2(a), reinjection after holding, one-more wall is formed within the C_{60} microtubes as shown in Fig. 2(b). When the 2-propanol was injected in stepwise mode as shown in Fig. 3(a), complex inner structures were formed including small microtubes on the top of large microtubes and porous inner structures. The changes in inner structure of vertically aligned C_{60} microtubes by modification of injection mode give a possibility that the inner structures are mud-like phases, while the outer surfaces are rigid solids.

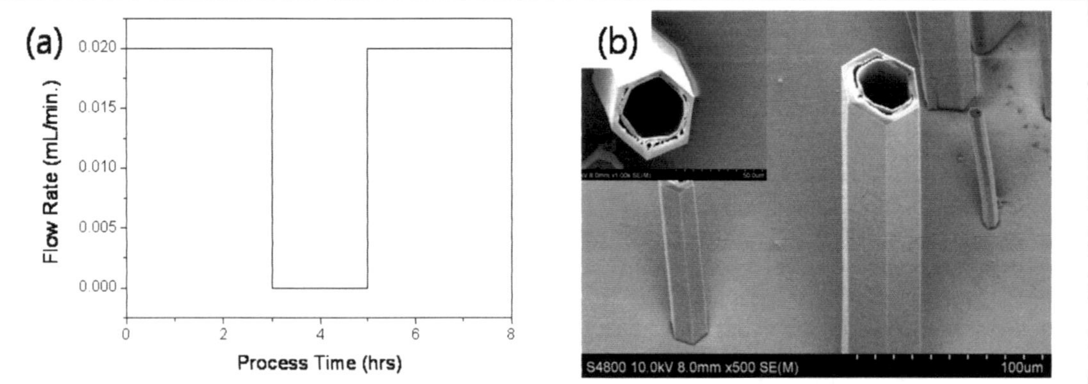

Figure 2. SEM morphology of vertically aligned C_{60} microtubes with double walled inner structure. (a) 2-propanol injection mode and (b) double walled C_{60} microtubes.

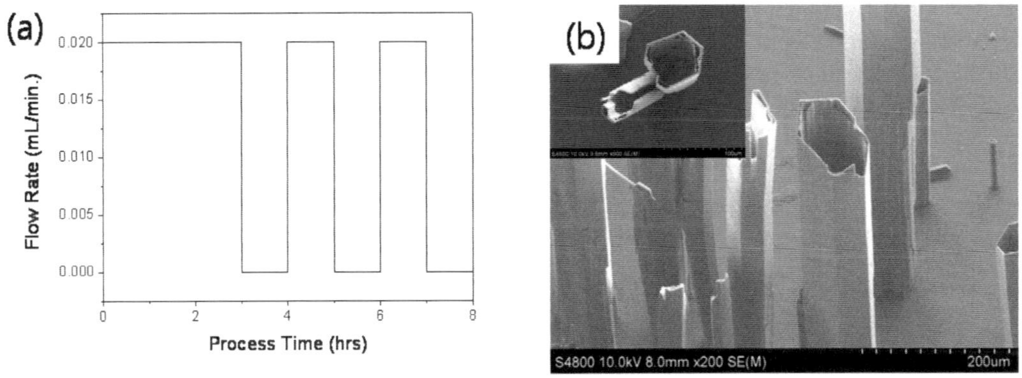

Figure 3. SEM morphology of vertically aligned C_{60} microtubes with complex inner structures including porous inner walls and small tubes on the wall of large tubes. (a) 2-propanol injection mode and (b) double walled C_{60} microtubes.

3.2. Effect of addition of ferrocene doped C_{60} on the morphology of C_{60} crystals

In addition to the kinetic modification, the addition of defects by addition of ferrocene doped C_{60} changes the morphology of C_{60} crystals dramatically as shown in Fig. 4. The tree-shaped structures with many branches composing of bundles of thin rods can be fabricated on the porous alumina substrates.

Figure 4. SEM micrographs of C_{60} trees on the porous alumina substrate prepared from the mixture of C_{60} and ferrocene doped C_{60} by 6 mol. %. (a) Low magnification micrograph showing the forest of C_{60} trees and (b) closer look of C_{60} tree with branches. (c) Closer image near the branches and (d) high magnification of end of branches showing bundles of needle shaped C_{60} crystals.

One interesting feature is that the structure is not affected by the modification of 2-propanol injection rate in the range from 0.01ml/min. to 0.04ml/min. It means that the structure is deeply dependent on the including crystal defect induced by addition of ferrocene doped C_{60}, rather than kinetic parameters.

The TEM observation shown in Fig. 5(a) gives a factor that each rod forming the bundle is rolled form of thin plate. The thin plates are divided into thin needles with sharp tips as shown in Fig. 5(b). The HRTEM image and diffraction patterns in Fig. 5(c) and (d) shows the thin plates are well crystallized. These structural characters give a possibility that they are formed by splitting and deformation due to the lattice strain due to inclusion of ferrocene doped C_{60}. However, to find out the detail mechanisms and structure, more analysis is required.

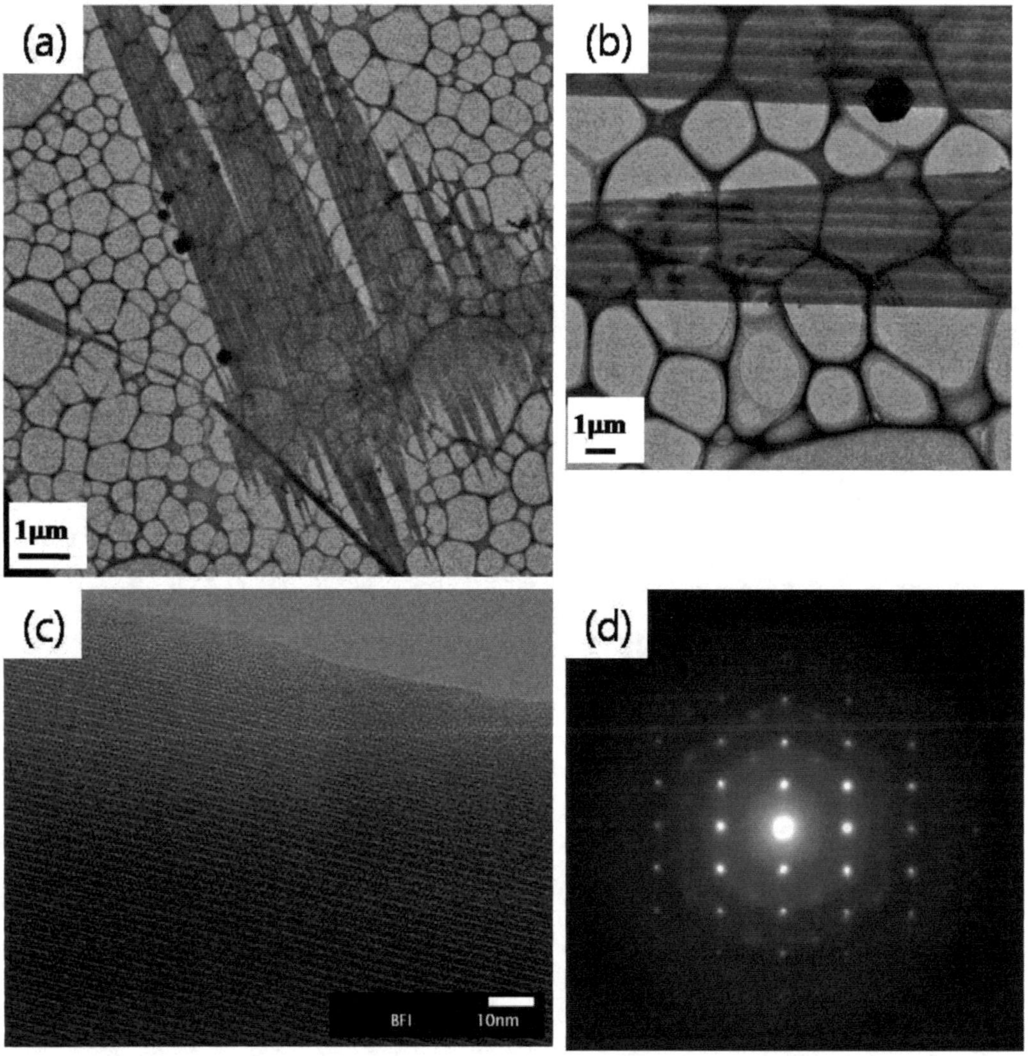

Figure 5. TEM micrographs of C_{60} trees prepared from the mixture of C_{60} and ferrocene doped C_{60} by 6 mol. %. (a) Low magnification image, (b) high magnification image showing splitting, (c) high resolution TEM image and (d) diffraction pattern.

4. Summary

The morphology of vertically aligned C_{60} microtubes is modified by controlling kinetic factors and including of lattice defects. The controlling kinetic factors by modified injection mode of 2-propanol gives various inner structure of vertically aligned C_{60} microtubes such as filled inner structure, double walled inner structure and porous inner structures. The addition of lattice defects by inclusion of ferrocene doped C_{60} gives tree-shaped C_{60} crystals with many branches consisting of bundles of thin needles.

References

[1] Kroto H W, Heath J R, O'Brien S C, Curl R F and Smalley R E 1985 *Nature* **318** 162
[2] Krätschmer W, Lamb L D, Fostiropoulos K and Huffman D R 1990 *Nature* **347** 354
[3] Dresselhaus M S, Dresselhaus G and Eklund P C 1996 *Science of Fullerenes and Carbon Nanotubes* (San Diego: Academic Press).
[4] Miyazawa K, Kuwasaki Y, Obayashi A and Kuwabara M 2002 *J. Mater.Res.* **17** 83
[5] Miyazawa K, Hamamoto K, Nagata S and Suga T 2003 *J. Mater. Res.* **18** 1096
[6] Minato J and Miyazawa K *Carbon* 2005 **43** 2837
[7] Tachibana M, Kobayashi K, Uchida T, Kojima K, Tanimura M and Miyazawa K 2003 *Chem. Phys. Lett.* **374** 279
[8] Sathish M, Miyazawa K and Sasaki T 2007 *Chem. Mater.* **19** 2398
[9] Cha S I, Miyazawa K and Kim J-D 2008 *Chem. Mater.* **20** 1667

Growth mechanism of vertically aligned fullerene microtubes prepared by the liquid-liquid interfacial precipitation method

S Toita[1], K Miyazawa[2], K Hotta[2] and M Tachibana[1]

[1]Yokohama City University, Yokohama, 236-0027, Japan
[2]National Institute for Materials Science, Tsukuba, 305-0044, Japan

E-mail: toita.sho@nims.go.jp

Abstract. C_{60}-based materials such as FNWs and FNTs are new candidates for organic electronics and solar cells. Recently, a method to produce vertically-aligned fullerene microtube crystals (VFMTs) has been established. The VFMTs are exepected to be used as field emission devices, solar cells, fuel cells, and so forth. To realize such attractive applications, structures and arrangements of VFMTs should be controlled. In oder to know the growth mechanism of VFMTs, Scanning electron microscopy (SEM) observations were performed for several injection periods of VFMTs. In the initial stage of synthesis (IPA injection time of 60 min), vertically aligned thin non-tubular fullerene microwhiskers (VFMWs) formed on AAO substrates. The outer and inner diameter of VFMWs became increased with increasing the injected amount of IPA. And also the size of tube holes became larger. The tubular structures appeared at an injection time of around 90 min. At an injection time of 120 min, vertically aligned FMTs with hexagonal cross sections and holes were obtained and their diameter and structure did not change up to the final stage of injection time of 300 min.

1. Introduction

Since C_{60} fullurene works as an efficient electron acceptor and has a relatively high electron mobility, C_{60}-based materials are expected as promising candidates in fields of organic electronics and solar cells. It is known that C_{60} forms quasi one-dimensional single crystals such as FNWs (fullerene nanowhiskers [1,3]), tubular fullerene nanowhiskers (fullerene nanotubes: FNTs [2]) and relatively short fullerene nanowhiskers (fullerene nanorods [4]). "Fullerene nanorods" is an alias for short fullerene nanowhiskers. These materials are obtained by the liquid-liquid interfacial precipitation (LLIP) method [3]. An addition of isopropyl alcohol (IPA) to a C_{60}-saturated toluene solution induces the nucleation and growth of C_{60} nano and micro whiskers. The order of addition can be changed in the LLIP method [7]. When we use pyridine in place of toluene, the products become C_{60} nanotubes.

Somani et al. performed organic thick film solar cell using short C_{60} nanowhiskers. To make more efficient organic solar cells (OSCs) using C_{60} materials, vertically aligned C_{60} whiskers must be more appropriate. Recently, Cha et al. succeeded to produce vertically aligned C_{60} microtubes (VFMTs) by modifying the LLIP method shown in ref. 3. In the method by Cha et al., an anodized alumina (AAO) membrane is set between the upper C_{60}-saturated toluene solution and the lower IPA. When IPA is injected into the C_{60}-toluene solution through the membrane from the lower side, V-FMTs form on the AAO substrates. C_{60}-based materials with designed structures such as VFMTs are expected to be used in various applications. For example, field emission devices, solar cells, fuel cells, and so on. But those applications can't be realized without the design control of VFMTs and the detailed growth mechanism of VFMTs has to be clarified.

To understand the growth mechanism of VFMTs, we performed scanning electron microsopy (SEM) observation for the samples of several injection periods (injection time 60, 90, 120, 180 and 300 min).

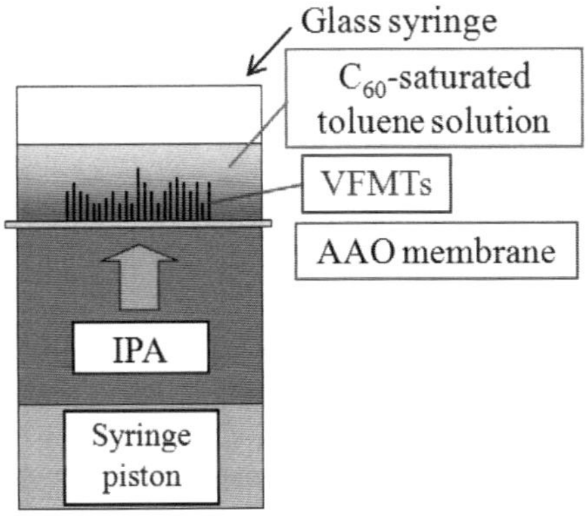

Figure 1. Schematic of synthesizing appratus for VFMTs.

2. Experimental

A schematic of synthesizing apparatus for VFMTs is shown in Fig.1. 3 mL of C_{60}-saturated toluene solution in a glass syringe was kept at 5 °C in an incubator. Then IPA was injected at a rate of 0.02 mL/min into the toluene solution through the membrane. Completely grown VFMTs were obtained when 300 min passed from the beginning of IPA injection. In this case, 6 mL of IPA was totally injected. Four injection periods of 60, 90, 120, and 180 min were adopted to observe the growth process of VFMTs. The amount of injected IPA for each injection time is shown in Table.1.

The sample morphology was observed by a scanning electron microscope (JSM-6700F by JEOL). 100 samples were analyzed to obtain the average values of diameter, area of cross section and walls thickness.

The cross-sectional polygonal plane of a VFMT is approximated by a circle whose area is the same with the polygonal area. The outer and inner diameters of the VFMTs are calculated from the area of this circle. The wall thickness of VFMTs is calculated from the inner and outer diameters thus obtained.

Table.1 Synthesis conditions for each injection period.

Injection time(min)	**C_{60}-toluene solution (mL)**	**Injected IPA (mL)**	**Total solution volume of systems**
60	3.0	1.2	4.2
90	3.0	1.8	4.8
120	3.0	2.4	5.4
180	3.0	3.6	6.6
300	3.0	6.0	9.0

3. Results and discussion

SEM images for samples in the initial injection period (injection time 60 min) are shown in Fig.2. Vertically alinged thin fullurene micro non-tubular whiskers formed on a AAO membrane. All the prepared samples exhibited various morphologies with different diameters and lengths.

The typical cross-sectional shapes of VFMWs are shown in Figs.2(a)-(c). In the obseved 100 samples, about 61 VFMWs had no holes on their top surface like Fig.3(a). 13 of the VFMWs had small holes as shown in Fig.2 (b). The diameter of this type of VFMWs is less than 300 nm. The other VFMWs had larger holes (500 nm ~ 2.62 μm) (Fig.2(c)). The classification criteria for the morphological types of VFMWs are shown in Table.2.

Table 2. Classification criteria for the morphological types of VFMWs for the injection time of 60 min.

	Type (a)	**Type (b)**	**Type (c)**	**Others**
Hole shape approximation	No hole	Circle	Circle	-
Inner diameter	0 μm	~ 0.3 μm	0.3 ~ 2 μm	-

The "Abundance of sample type [%]" is defined as follows.

$$\text{Abundance of sample type} = \frac{\text{The number of the same type samples}}{\text{The total number of observed samples}} \times 100 \ [\%]$$

The ratio of hole area to the cross-sectional area in VFMWs is calculated as follws.

$$\text{Hole area ratio} = \frac{\text{Area of hole } [\mu\text{m}^2]}{\text{Area of corss section } [\mu\text{m}^2]} \times 100 \ [\%]$$

The average percentage of hole area was only 1.49 % of the total cross-sectional area of VFMWs. This means that most of the VFMTs formed in the initial injection period have the non-tubular morphology. The average diameter at an injection time of 60 min was 6.15 μm.

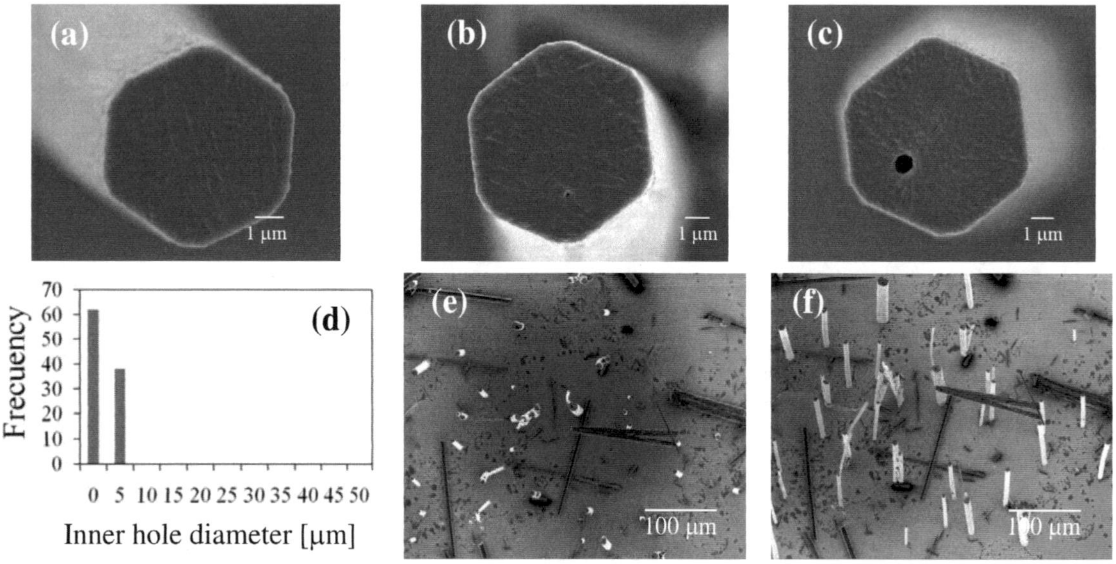

Figure 2. SEM images of VFMWs (injection time 60 min). (a)-(c) typical cross sections of VFMWs for an injection time of 60 min. (a) VFMW without hole, (b) VFMW with a small hole, (c) VFMW with a relatively large hole. (d) Inner hole diameter distribution of VFMWs. (e) Plan view and (f) inclined view of VFMWs by SEM.

The diamter of VFMWs increased with increasing the amount of injected IPA. The tubular structures appeared at an injection time of around 90 min for diameters greater than a certain value (Fig.3). All the tubes showed regular hexagons, but their tube holes were not hexagonal. The holes can be classified into 3 types of morphologies. 24 % of the VFMTs had such irregular shapes as shown in Fig.3(a), where various shapes of holes like circular, trigonal, pentagonal forms and so forth were observed. And their ratio of hole area against the area of cross-section was less than 10 %.

The most typical tube structure is shown in Fig.3(b). The VFMTs show the incomplete hexagonal holes with surrounding terraces. The average inner diameter of VFMTs like Fig.3(b) was less than 10 μm. 10 % of them had relatively large holes whose inner diamter was more than 10 μm (Fig.3(c)) . Non-tubular whiskes like Fig.2(a) or VFMWs with small circular holes like Fig.2(b) are also observed. There is a correlation between the value of inner diameter and the hole shape. The more inner diameter becomes larger, the more hole shape becomes closer to hexagonal. This correlation was also observed in the whole injection periods. The classification criteria for the morphological types for the injection time of 90 min are shown in Table.3.

Table 3. Classification criteria for the morphological types of VFMTs for the injection time of 90 min.

	Type (a)	**Type (b)**	**Type (c)**	**Others**
Hole shape approximation	Irregularly shaped polygons	Imperfect hexagon	Imperfect hexagon	No hole or small circular hole
Inner diameter	2 ~ 5 μm	5 ~ 10 μm	10 ~ 20 μm	0 ~ 2 μm

In this injection period, the average ratio of hole area to the cross-sectional area was about 17 % and the average diameter was 15.7 μm.

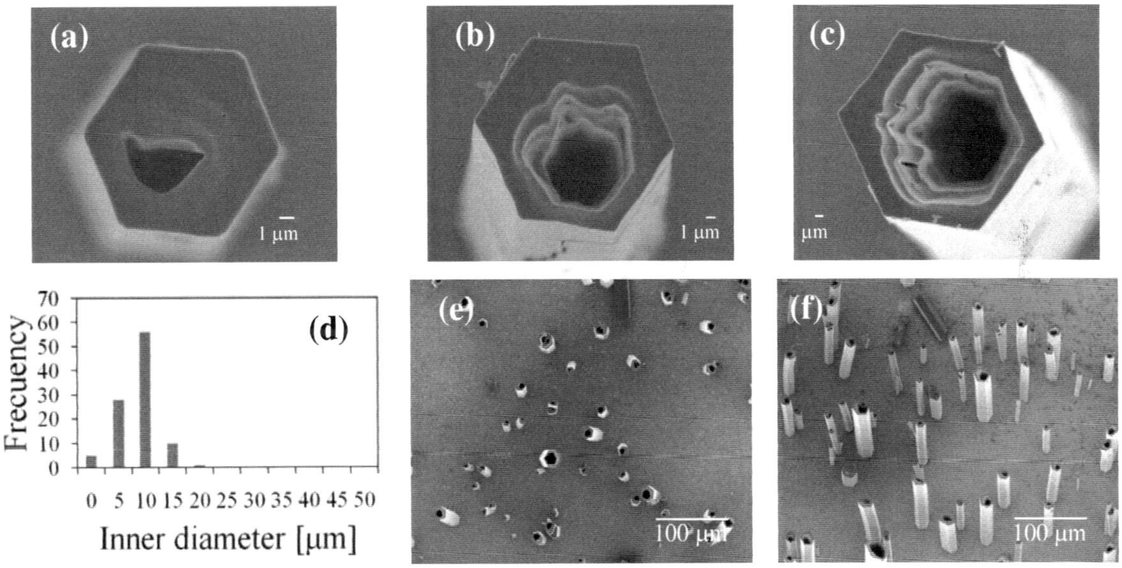

Figure 3. Typical cross-sectional SEM images (a, b, c, e, f) of VFMTs for the injection time of 90 min (a) VFMT with an irregularly shaped hole, (b) VFMT with an imperfect hexagonal hole and (c) VFMT with relatively large hole. (d) Inner diameter distribution of VFMTs. (e) Plan view and (f) inclined view of the VFMTs by SEM.

The VFMTs with hexagonal cross sections and holes were obtained for an injection time of 120 min. These structures were almost the same as the final products of the injection time of 300 min (Fig.5). In this injection period, 92 % of them had the hexagonal cross sections and tube holes as shown in Figs.4(a) and (b). But the rest of 5 % had imperfect hexagonal holes with wavy sides like Fig.4(c).

In the sample as shown in Fig.4(b), the shape of hole has a hexagonal symmetry and the tube wall is thin, while the sample shown in Fig.4(a) has a thicker tube wall without the hexagonal symmetry. It is found that the shape of holes approaches regular hexagons with decreasing the wall thickness. The classification criteria for morphological types for the injection time of 120 min are shown in Table.4.

Table 4. Classification criteria for the morphological types of VFMTs for the injection time of 120 min.

	Type (a)	Type (b)	Others
Hole shape approximation	Incomplete hexagon	Regular hexagon	Other polygons or irregularly shaped holes
Wall thickness	4.5 μm ~ 5.5 μm	3.5 ~ 4.5 μm	5.5 ~ 6.5 μm

The average ratio of hole area to the cross-sectional area was about 44 %. And the average outer diameters for the injection time of 120 min and 180 min were 28.4 μm and 28.8 μm, respectively.

No significant difference was observed between the samples for the injection time of 120 min and those for the injection time of 180 min.

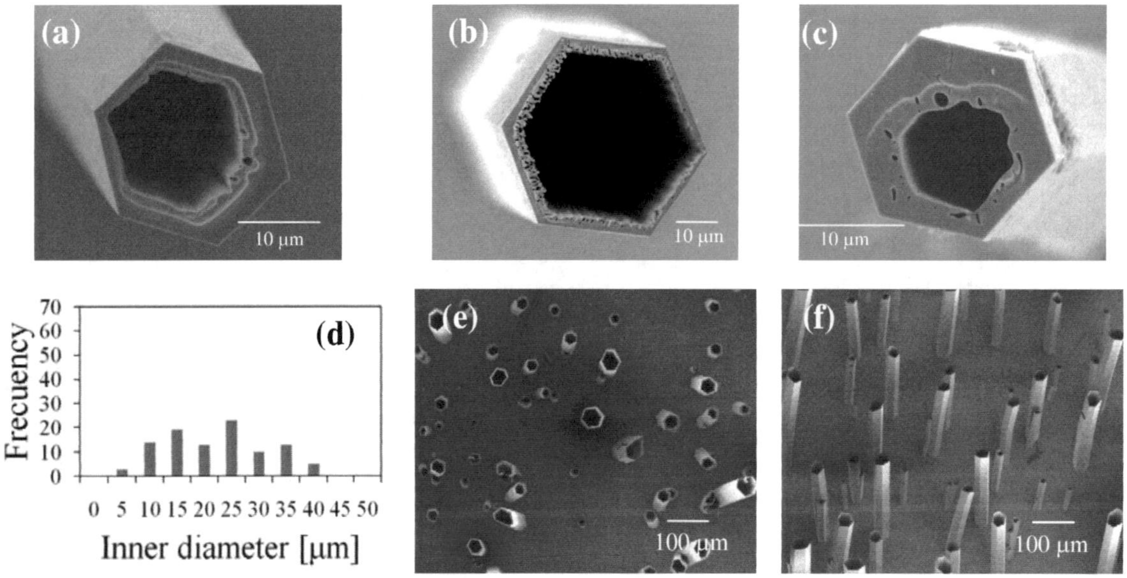

Figure 4. Typical cross-sectional SEM images (a, b, c, e, f) of VFMTs for the injection time of 120 min (a) VFMT with an incomplete hexagonal hole, (b) VFMT with a regular hexagonal hole and (c) an example for other polygonal holes. (d) Inner diameter distribution of VFMTs. (e) Plan view and (f) inclined view of VFMTs by SEM.

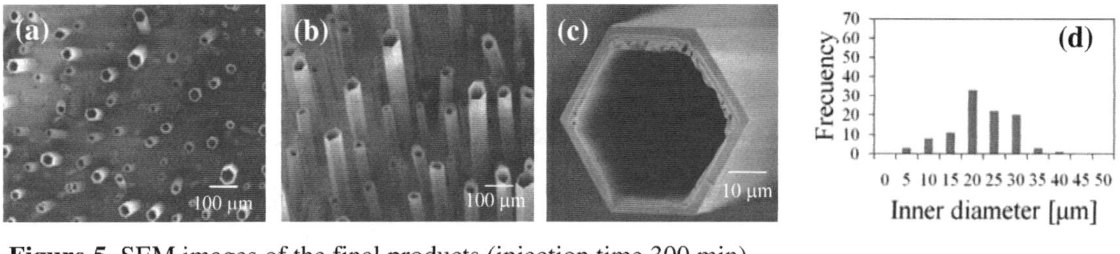

Figure 5. SEM images of the final products (injection time 300 min).
(a) Plan view and (b) inclined view of VFMTs. (c) Cross-sectional view of a VFMT. (d) Inner diameter distribution of VFMTs for the injected time 300 min.

Fig.6 shows the distribution of outer diameter of VFMTs for the different injection periods. The median of outer diameter for each period shifts to right with increasing the injected amount of IPA. For the injection time of 120 min, the distribution of outer diameter became broader and showed the similar curves. The distribution of inner diameter of VFMTs also showed the same behaviour.

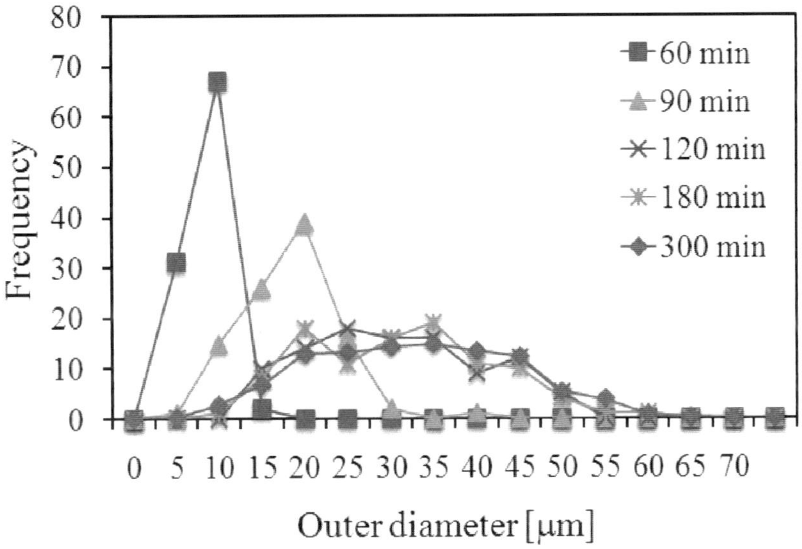

Figure 6. Outer diameter distribution of VFMTs measured as a function of injection time.

4. Conclusion

It has been found that the vertically-aligned fullerene microtubes exhibit solid non-tubular structures in their initial injection period. At the injection time of 60 min, vertically aligned thin non-tubular C_{60} whiskers formed on the AAO substrates. The ratio of the hole area to the cross-sectional area and the outer and inner diameters of the VFMWs increased with increasing the injection time.

The tubular structures appeared at the injection time of around 90 min. At the injection time of 120 min, vertically aligned FMTs with hexagonal cross sections were obtained and their morphology did not change up to the final stage of injection time of 300 min. This result shows that the formation of VFMTs completes within the injection time of 120 min.

References

[1] Miyazawa K, Obayashi A and Kuwabara M, 2001 *J. Am. Ceram. Soc.* **84** 3037
[2] Miyazawa K, Minato J, Yoshii T, Fujino M and Suga T, 2005 *J. Mater. Res.* **20** 688
[3] Miyazawa K, Kuwasaki Y, Obayashi A and Kuwabara M, *J. Mater. Res.* **17** (2002) 83
[4] Jin Y, Cury R J, Sloan J, Hatton R A, Chong L C, Blanchard N, Stolojan V, Kroto H W, and Silva S R P, 2006 *J. Mater. Chem.* **16** 3715
[5] Somani P R, Somani S P, and Umeno M, 2007 *App.phys.lett* **91** 173503
[6] Cha S I, Miyazawa K and Kim J D 2008 *Chem. Mater.* **20** 1667
[7] Miyazawa K, Minato J, Mashino T, Yoshii T, Kizuka T, Kato R, Tachibana M and Suga T 2005Characterization of the liquid-phase synthesized fullerene nanotubes and nanowhiskers *Proc. the 2nd JSME/ASME International Conference on Materials and Processing 2005 - The 13th JSME Materials and Processing Conference- M&P2005, (Seattle, USA, 19-22 June 2005)* pp (SMS23)-1 - (SMS23)- 4

Adsorption characteristics of heat-treated fullerene nano-whiskers

Z. -M. Wang[1], R. Kato[2], K. Hotta[2] and K. Miyazawa[2]

[1]Energy Storage Materials Group, Energy Technology Research Institute, National Institute of Advanced Industrial Science and Technology, 16-1 Onogawa, Tsukuba, Ibaraki 305-8569, Japan

[2]Fullerene Engineering Group, Advanced Nano Materials Laboratory, National Institute for Materials Science, 1-1 Namiki, Tsukuba, Ibaraki 305-0044, Japan

E-mail: zm-wang@aist.go.jp

Abstract. Fullerene nanowhiskers (FNWs) were synthesized by the liquid-liquid interfacial precipitation method and the adsorption properties of their heat-treated samples were characterized. It was found that vacuum-annealed FNWs at a high temperature are of microporous materials and, especially, ultramicropores are highly developed in these materials. Porosities even remain in samples after heat treatment at a temperature higher than 2273 K. The presence of ultramicroporosity is indicative of the molecular sieving properties of the vacuum-annealed FNW materials, suggesting the possibilities of their application as new materials for gas separation and gas storage.

1. Introduction

Fullerene (or C_{60} molecule) has attracted a great deal of concerns since its finding in 1985 [1] from both fundamental and industrial fields, because of its unique structure and properties, and hopeful applications as superhard materials, lubricants, composite components, solar cell materials, and so forth. It is well-known that solid fullerene forms molecular crystal via van der Waals-type interaction among fullerene molecules. The fullerene crystal is generally of a face-centered cubic (fcc) structure with a lattice constant of 1.417 nm and C_{60}-C_{60} distance of 1.002 nm, sublimates at a lower temperature (523~573 K) [2, 3], and is of semiconductor property with a band gap of 1.7 eV [4]. Sometimes the morphology of fullerene crystal can be controlled by controlling the crystalline growth conditions. Michaud and Céolin have found that needle-like fullerene crystals can be obtained by slow evaporation of solvents from the fullerene solutions in dichloromethane, n-hexane, etc [5, 6]. However, these needle-like crystals generally have big (20 to 100 μm in diameter) and long (0.2 to 20 mm in length) sizes.

Recently, Miyazawa et al have found that fullerene nanowhisker (FNW) and nanotube (FNT) consisting of C_{60} molecules with nanosized diameter and high aspect ratio can be simply obtained by liquid-liquid interfacial precipitation method [7-9]. In this method, a poor solvent of C_{60} molecules is contacted with a good solvent of C_{60} molecules and the interfacial surface between the two solvents serves as the nucleation sites for fibrous crystal growth of C_{60} molecules which are previously dissolved in the good solvent. While the FNW and FNT materials are of attractive structures and properties, the heat-treated FNWs show unusual light-emitting property [10] and are expected as a new group of functional materials. In this study, we applied gas/vapor adsorption techniques to investigate the structural (porous) and surface properties of the heat-treated FNWs.

2. Experimental

2.1. Materials

Isopropyl alcohol (IPA) and toluene were used, respectively, as the poor and good solvents of C_{60} for synthesizing the FNW samples. C_{60} with purity of 99.5% purchased from MTR Ltd. (Cleveland, OH) was first dissolved into toluene to form a saturated solution containing 0.3 mass% C_{60} molecules and IPA was gently layered on the solution to form a liquid-liquid interface. The solutions were kept still at 278 K for 7 days. The obtained needle-like C_{60} crystals, i.e., fullerene nanowhiskers (FNW) were dried at room temperature (RT).

The FNW powders were evacuated at RT in a quartz cell, which was sealed by an oxyhydrogen flame in a vacuum with a pressure less than 10^{-5} Pa. The vacuum-sealed FNW powders were first heat-treated in a muffle furnace at 1173 K for 1 h. The collected 1173 K-treated samples were further treated in a carbon furnace under argon atmosphere at 2273 and 2773 K for 1 h. The obtained samples are denoted as FNW-T where T stands for the heat treated temperature.

2.2. Characterization and adsorption

The microscopic structures of samples were examined using a field-emission scanning electron microscope (SEM, JEOL JSM-6700F) and a high-resolution transmission electron microscope (HRTEM, JEM-4010, JEOL).

Adsorptions on samples were measured at 77 K for N_2 and 298 K for H_2O and C_6H_{14} by a commercial volumetric apparatus (BELSORP 18A). The equilibrium time for each adsorption point was maintained at 10 min. Before N_2 adsorption, samples were pre-evacuated under 1 mPa for removing impurities. Any strong specific interaction between FNW-T samples and N_2, H_2O, or hexane molecule is hardly considered, since the FNW-T samples contain almost no surface functional sites. Hence, the FNW-T powders after N_2 adsorption were reused for H_2O adsorption and those for H_2O adsorption reused for hexane adsorption. The used FNW-T samples were evacuated again at 393 K for 2 h before reusing for the next run adsorption.

3. Results and Discussion

3.1. Microscopic structures

Miyazawa et al. has reported that structure and size of needle-like C_{60} crystals can be controlled by changing the kind and volume ratio of solvents. While pyridine solvent leads to formation of fibrous fullerene crystals with tubular structure, toluene and *m*-xylene solvents give rise to solid fullerene nanowhiskers [8, 9, 11]. Figure 1 shows the FE-SEM and TEM images of FNWs synthesized in this work. Coinciding with the previous reports, FNWs consist of large number of needle-like particles with a diameter ranging from 300 through 700 nm and a length generally surpassing 5 μm. The TEM image clearly confirms that the needle-like particles are of solid property rather than tubular structure. The previous research reported the powder of pristine C_{60} with a very small specific surface area (~20 m^2/g), despite that pores are directly detected on the surface of the FNWs prepared by use of benzene and IPA [12]. The solvent-wetted FNWs are of a hexagonal structure but transformed to the face-centered cubic (fcc) structure when completely dried, which is similar to that of the bulk C_{60} powders, however, containing lots of lattice defects [13].

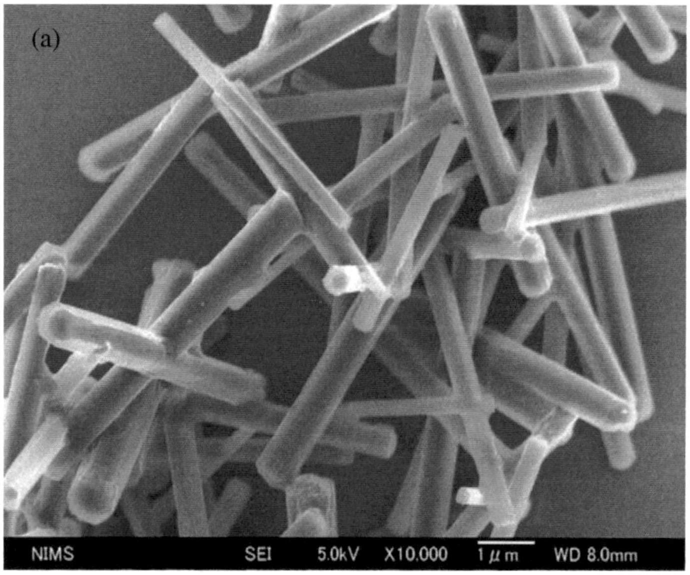

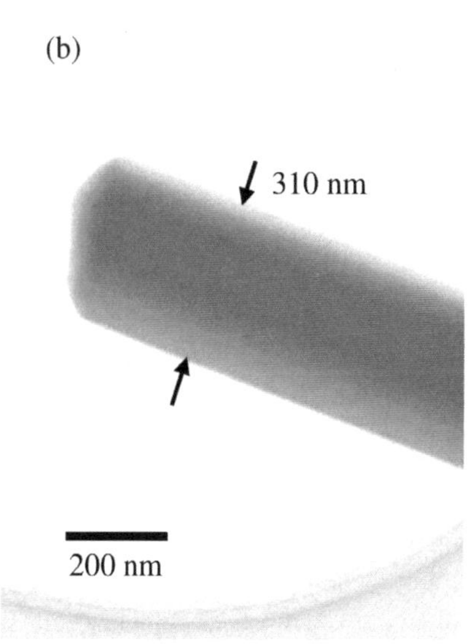

Figure 1. (a) FE-SEM and (b) TEM pictures of the pristine FNWs.

On the other hand, the morphologies of the heat-treated FNWs are not greatly changed as compared to the pristine FNWs (microscopic data omitted for simplicity). High-resolution TEM observation indicates that the fullerene crystal structure is completely destroyed by heat treatment above 1173 K and glassy carbon-like structures are formed, with the stacking carbon hexagon layers increasing with the increase of heat treatment temperature [14].

3.2. Nitrogen adsorption

Figure 2 shows the N_2 adsorption isotherms at 77 K on FNW-T. The N_2 adsorption on FNW-1173 presents a sharp uprising at very small relative pressure (P/P_0) ranges, followed by a gradual increase at higher P/P_0 ranges, indicating that the material possibly contains micropores and possesses comparatively a large part of external surface [15]. Hysteresis phenomenon is observed for this material (from the desorption branch of the isotherm), which is extended to very low P/P_0 ranges (<0.2). Since chemisorption or any kind of strong specific adsorption is not possible for N_2 adsorption, and structural changes such as swelling or shrinking of the material are hardly realized due to N_2 adsorption, the low pressure adsorption hysteresis of N_2 adsorption in this case can most possibly be attributed to the existence of pores or pore apertures with smaller

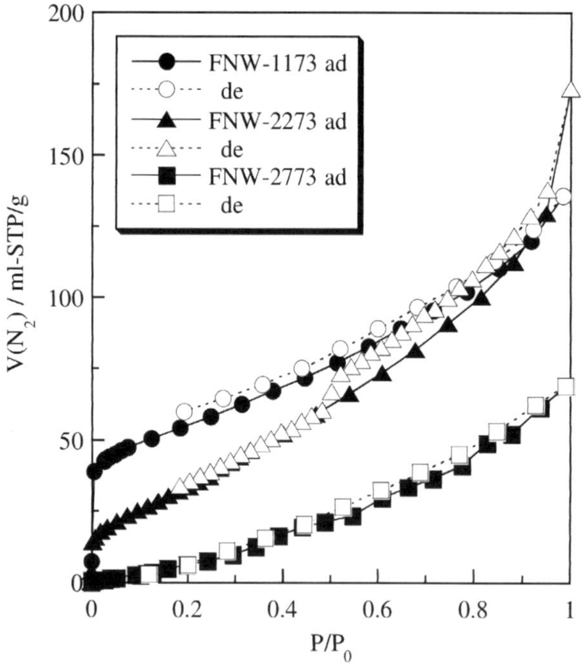

Figure 2. N_2 adsorption isotherms at 77 K on FNW-

or competitive sizes as compared to nitrogen molecules.

N_2 adsorption gradually decreases with the increase in the heat treatment temperature. However, the adsorption isotherm on FNT-2273 is still of a typical type IV shape [15], indicative of formation of mesoporosity in this material. Thus, porosities in FNW-T are very tolerant to high temperature. On the contrary, the adsorption isotherm on FNT-2773 changes to a typical type III shape, manifesting the formation of low affinity surface toward N_2 molecules and almost disappearance of porosity.

For relative comparison among the heat-treated FNW samples, the following well-known Brunauer-Emmett-Teller (BET) equation was applied to obtain the pore parameters of the samples (though it is argument to use it for pore analysis of microporous materials):

$$n = n_m Cx/(1-x)/(1-x+Cx) \quad (1)$$

where n is the adsorption amount at $x = P/P_0$, n_m is the monolayer capacity of adsorption from which specific surface area, a_{BET}, can be calculated, and C is a constant reflecting the interaction energy between adsorbate molecules and adsorbent surface. Table 1 shows the values of a_{BET} and C for the three samples from N_2 adsorption. The a_{BET} values of the FNW-T samples are much greater than that of the pristine FNW (around 20 m^2/g). Especially, the greatest a_{BET} value of FNW-1173 implies the evolution of a great amount of pores by vacuum annealing of FNW, lots of which are even retained after heat treatment at 2273 K. The greatest C value of this sample relates to the sharper uprising in N_2 adsorption at small P/P_0 ranges, which comes from a strong adsorbate-adsorbent interaction. The porosity of FNW-1173 may have some relationship with the destruction or partial destruction of fullerene cages, by which pore openings with molecular sizes are possibly formed.

3.3. H_2O and hexane adsorption

Figure 3 shows the H_2O adsorption isotherms at 298 K on FNW-T samples. While the adsorption amounts of H_2O on the three samples are the same sequence as those of N_2 adsorption (Figure 2), the adsorption isotherm of H_2O on FNW-1173 presents an unusual shape. It was known that H_2O adsorption on porous carbonaceous materials is generally very low at low P/P_0 ranges because of the hydrophobic carbon surface, and suddenly increases when P/P_0 surpasses a medium value (above ~0.4) [15]. An S-shaped isotherm of H_2O, characteristic of an evident uprising in H_2O adsorption from the low P/P_0 ranges, can be observed on ultramicroporous materials due to the enhanced potential fields from the opposite walls of the ultrafine pores [15]. Although the specific surface area of FNW-1173 is much smaller as compared to the ordinary porous carbons of high specific surface area such as activated carbons, the H_2O adsorption on this material is sufficiently high at the low P/P_0 ranges so that the S-shape of the isotherms rather becomes evident at the low P/P_0 ranges. This result strongly suggests the existence of ultramicropores in the material which induces stronger interaction between pore walls and small molecules such as H_2O.

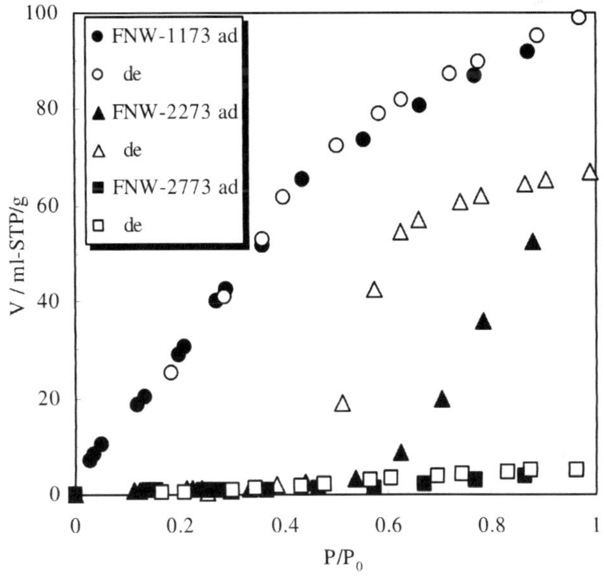

Figure 3. H_2O adsorption isotherms at 298 K on FNW-T samples.

To the contrast, the H_2O adsorption on FNW-2273 is very low at P/P_0 < 0.5, characteristic of typical carbonaceous materials, with a broad hysteresis branch due to the mesoporosity. Because of less porosity, H_2O adsorption on FNW-2773 is very small in the whole P/P_0 range.

The values of BET specific surface area were calculated from H_2O adsorption isotherms and shown also in Table 1. Although the a_{BET} values from H_2O adsorption are much smaller in comparison with that from N_2 adsorption for the FNW-2273 and -2773 samples because of the depressed H_2O adsorption at small P/P_0 ranges, that from H_2O adsorption is comparable with that from N_2 adsorption for the FNW-1173 sample, due to the enhanced H_2O adsorption at very small P/P_0 ranges.

Figure 4 shows the hexane adsorption isotherms at 298 K on FNW-T samples. All the isotherms exhibit a slight uprising at small P/P_0 ranges and, subsequently, a gradual increase at the higher P/P_0 ranges. The characteristics of isotherms on FNW-1173 and -2273 are the great low pressure hysteresis branches, indicating that both materials contain micropores with very small pore widths. Especially, the FNW-1173 material demonstrates a greater hysteresis; its hexane adsorption is comparable with that on FNW-2273 in the whole range despite that its N_2 and H_2O adsorptions are much higher. This result suggests that a part of pores in the material are hardly accessible to hexane molecules which have a larger molecular size as shown in Table 1. Therefore, the material presents molecular sieve property, containing ultramicropores excluding molecules with sizes greater than 0.43 nm. The vacuum-annealed FNW materials can thus possibly be a group of new material for the purposes of gas separation and gas storage.

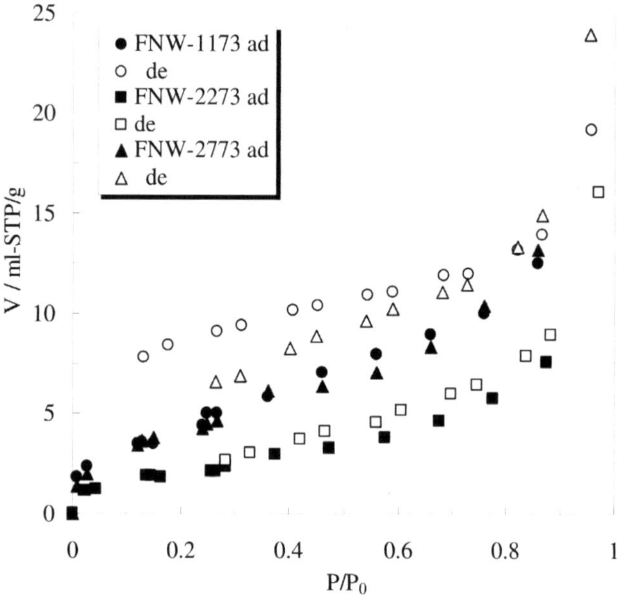

Figure 4. Hexane adsorption isotherms at 298 K on FNW-T samples.

Table 1. Pore parameters from BET analysis of adsorption isotherms.

Adsorbate	Molecular size / nm		Sample		
			FNW-1173	FNW-2273	FNW-2773
N_2	0.364	a_{BET} / m^2/g	195	140	80
		C	795	22	1.4
H_2O	0.265	a_{BET} / m^2/g	155	3	2
		C	4	26	33
Hexane	0.43	a_{BET} / m^2/g	50	45	25
		C	34	122	110

Journal of Physics: Conference Series **159** (2009) 012013 doi:10.1088/1742-6596/159/1/012013

4. Conclusion

Liquid-liquid interfacial precipitation using toluene and IPA solvents provides non-tubular FNWs with a diameter greater than 300 nm and a length longer than 5 μm. While the morphology of FNW is maintained by high temperature annealing, the fullerene structure is destroyed and changes to a glassy carbon-like structure. Adsorption measurements using probe molecules of different molecular sizes and of different polarities confirm that the vacuum-annealed FNW materials are highly porous and contain ultramicropores of molecular sizes, leading to molecular sieve properties. By optimizing the synthesis method and conditions, these materials can possibly become a group of new nanoporous materials for applications in gas separation and gas storage.

References

[1] Kroto H W, Heath J R, O'Brien S C, Curl R F and Smalley R E 1985 *Nature* **318** 162
[2] Fasel R, Agostino R G, Aebi P and Schlapbach L 1999 *Phys. Rev.* **60** 4517
[3] Bertoni G, Cepek C and Sancrotti M 2003 *Appl. Surf. Sci.* **212-213** 52
[4] Hirsch A and Brettreich M 2005 *Fullerene - Chemistry and Reactions* (Heppenheim:Wiley-VCH)
[5] Michaud F, Barrios M, Toscani S, López D O, Tamarit J L1, Agafonov V, Szwarc H and Céolin R 1998 *Phys. Rev. B* **57** 10351
[6] Céolin R, Tamarit J L1, López D O, Barrios M, Agafonov V, Allouchi H, Moussa F and Szwarc H 1999 *Chem. Phys. Lett.* **314** 21
[7] Miyazawa K, Obayashi A and Kuwabara M 2001 *J. Am. Ceram. Soc.* **84** 3037
[8] Miyazawa K, Kuwasaki Y, Obayashi A and Kuwabara M 2002 *J. Mater. Res.* **17** 83
[9] Minato J and Miyazawa K 2006 *Diamond and Related Materials* **15** 1151
[10] Asaka K, Kato R, Maezono Y, Yoshizaki R, Miyazawa K and Kizuka T 2006 *Appl. Phys. Lett.* **88** 051914
[11] Minato J and Miyazawa K 2005 *Carbon* **43** 2837
[12] Sathish M, Miyazawa K and Sasaki T 2007 *Chem. Mater.* **19** 2398
[13] Miyazawa K, Hamamoto K, Nagata S and Suga T 2003 *J. Mater. Res.* **18** 1096
[14] Kato R, Miyazawa K, Nishimura T and Wang Z M *J. Phy. Conf. Series* accepted.
[15] Rouquerol F, Rouquerol J and Sing K 1999 *Adsorption by powders & porous solids* (San Diego: Academic Press)

Carrier Injection Property of C_{60} Thin Film Transistor via Displacement Current Measurement

N Aoki[1], Y Chiba[1], S -R Chen[1], H Tsuji[1], M Ueno[1], K Ogawa[1], J P Bird[2] and Y Ochiai[1]

[1]Graduate School of Advanced Integration Science, Chiba University, Chiba, Japan
[2]Department of Electrical Engineering, University at Buffalo, SUNY, Buffalo, USA

E-mail: n-aoki@faculty.chiba-u.jp

Abstract. Carrier injection into a C_{60} thin film has been studied from characteristics of a field effect transistor and a displacement current measurement (DCM) using three kinds of metals, Mg, Ag, and Au. The Mg sample shows a highest field effect mobility of 0.62 cm^2/Vs and a lowest threshold voltage of 6 V among the samples. A carrier injection and the accumulation are confirmed in the DCM curves by sweeping voltage applied on the back gate. The electron number and the density injected from the electrodes are estimated from the integration of the DCM curves. However, only the curve of the Mg sample shows a two-stage transition in the DCM curves. It could be due to a diffusion of Mg atoms into the C_{60} layer and a transition of releasing electrons from the donor level could be observed at lower voltage region.

1. Introduction

Recently, organic semiconductors are getting applied for light emitting devices, and they are also expected to be applied for their drive transistors. However the operation mechanism, especially the origin of the carrier accumulation or injection, is not yet clear [1]. In *n*-type organic materials, C_{60} is known to show relatively high mobility [2, 3] therefore an appropriate metal electrode is desired to form a contact with a low injection barrier for further improvement of mobility for its application. Because of *n*-type FET, low-work-function metals are expected to decrease the injection barrier for C_{60} and to improve the mobility. For example, Eu, whose work function is smaller than the LUMO level of C_{60}, has been tried to form a FET [4]. We have, therefore, studied a carrier injection property of C_{60} thin film transistor using displacement current measurement (DCM), which has been used to characterize a carrier accumulation into an organic layer [5]. In this paper, we have confirmed carrier injection characteristics of three kinds of metal, Mg, Ag and Au, by the FET operation and the DCM with changing the measuring temperatures.

2. Sample and experimental setup

C_{60} thin films were deposited 100 nm by thermal evaporation on a thermally grown SiO_2 layer on a p^{++}-Si substrate which was used as a back gate. Three kinds of metals, Mg, Ag and Au, were used for the electrical contact and were deposited on the C_{60} thin films forming a bottom contact structure. In cases of the Mg and the Ag contact, Au thin film was deposited on the electrodes to avoid the oxidation. The samples were measured in vacuum of ~10^{-6} torr after annealing at 430 K for 24 hours. For a determination of the FET characteristic, current across the source and drain (I_{SD}) was measured with changing the back gate voltage (V_{bg}). For the DCM, displacement current (I_d) was measured between the electrodes on the top of the C_{60} layer (the source and the drain electrodes) and the back gate by sweeping the V_{bg} with a rate of 1 mV/s.

3. Results and discussion

V_{bg} dependences of the square root of I_{SD} are show in Fig. 1 with difference contact metals. All of the FETs show *n*- and enhancement type behaviour. The field effect mobility (μ), the threshold voltage and (V_{th}) of the DCM curve are shown in Table 1. The V_{th} in the Mg sample shows the smallest value of 6 V among these samples. It would be due to the smallest difference of energy between the LUMO level of C_{60} (3.6 eV) and the work function of the metal (3.6 eV for Mg, 4.3 eV for Ag, and 4.9 eV for Au). Consequently, the μ in Mg sample shows the highest value of 0.62 cm^2/Vs. It is confirmed that lowering of the barrier height at the interface of C_{60} and the metal is advantageous for increase of the mobility in a C_{60} based FET.

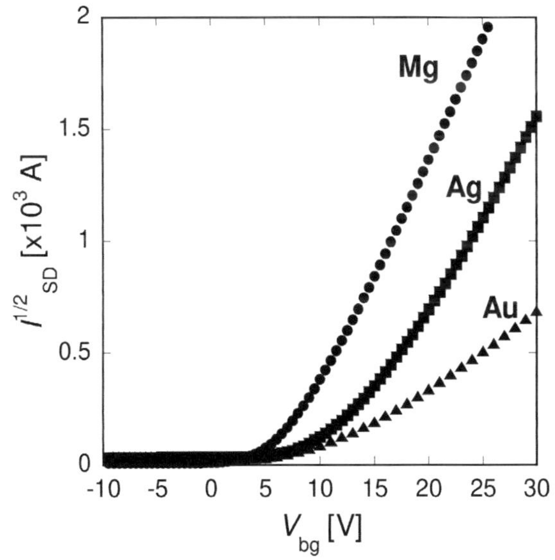

Figure 1. I_{SD} -V_{bg} characteristics of the FET structures with Mg (circle), Ag (square) and Au (triangle) as the contact metals. Square root of I_{SD} is used for the vertical axis to determine the V_{th}. The V_{th} in Mg, Ag and Au samples are 6, 10 and 11 V, respectively.

DCM curves in Mg, Ag and Au samples are shown in Figs. 2(a)-2(c), respectively. In the negative V_{bg} region, no carrier is injected into the C_{60} channel. Therefore, the C_{60} layer acts as an insulating layer and forms a series connection of the capacitances. The I_d can be considered to flow across the series capacitances of insulator layers of the C_{60} and the SiO_2 and would be expressed by the following equation,

$$I_d = (S_S + S_D)\frac{C_C C_S}{C_C + C_S}\frac{dV_{bg}}{dt}, \tag{1}$$

where S_S and S_D are the areas of the source and drain electrodes, respectively, and C_C and C_S are the capacitance of C_{60} and SiO_2 layers, respectively. The I_d can be calculated as 0.021 nA, which is indicated by broken line (*i*) in Fig. 2(a), from the practical values and it is very consistent with the I_d value at the left side of the DCM curve as shown in Fig. 2(a). On the other hand, when V_{bg} exceeds the V_{th}, the electrons begin to be injected from the metal electrodes and to be accumulated at the interface between the SiO_2 and C_{60} thin film. At enough positive V_{bg}, the I_d would be consequently expressed by the following equation,

$$I_d = S_C C_S \frac{dV_{bg}}{dt}, \tag{2}$$

where S_C is the areas of the C_{60} thin film. These scenarios are well applicable at the right hand of the DCM curve at enough positive V_{bg}. One of the features of these DCM curves of C_{60} thin films is a peak of the I_d just before the electron widely distributing over the C_{60} layer. The voltage which is The peak does not appear when the carriers are going to be depleted by sweeping the V_{bg} to the negative direction. Considering the sweep direction, it would be due to a difference of efficiency of the barrier shape for carrier injection at the interface. Because the carriers are injected into the C_{60} layer by tunnelling the barrier in a regime of Fowler-Nordheim-type field emission in the positive sweep [6]. A sudden increase of carrier injection could cause an excess current and show a peak in the DCM curve. Such a peak was observed also in a DCM curve of a pentacene FET [5]. On the other hand, thermionic emission would be dominant when the carrier is going back to the metal region. The carriers could be continuously returned to the metal region hence no peak would appear in this regime. The peak voltage (V_p) of the DCM curve are different in each sample as shown in Table 1. Nevertheless, it can be confirmed that a trend of the voltage shift of the V_p is similar to that of the V_{th}.

The injected carrier (electron) number and the density can be estimated by integrating the DCM curve,

$$Q = \int_{t_0}^{t_1} \{I_d(t) - I_{d0}\} dt, \tag{3}$$

where I_{d0} is displacement current before starting the injection of carriers into the C_{60} layer, and t_0 and t_1 are the times when the slope of triangle wave of V_{bg} is inverted. For the Mg sample, the amount of the injected carrier and the concentration were estimated 3.5×10^{-8} C and 1.6×10^{12} /cm^2, respectively, by integrating the area colored by gray in Fig. 2(a). In the same way, the values in Ag and Au samples are also summarised in Table 1.

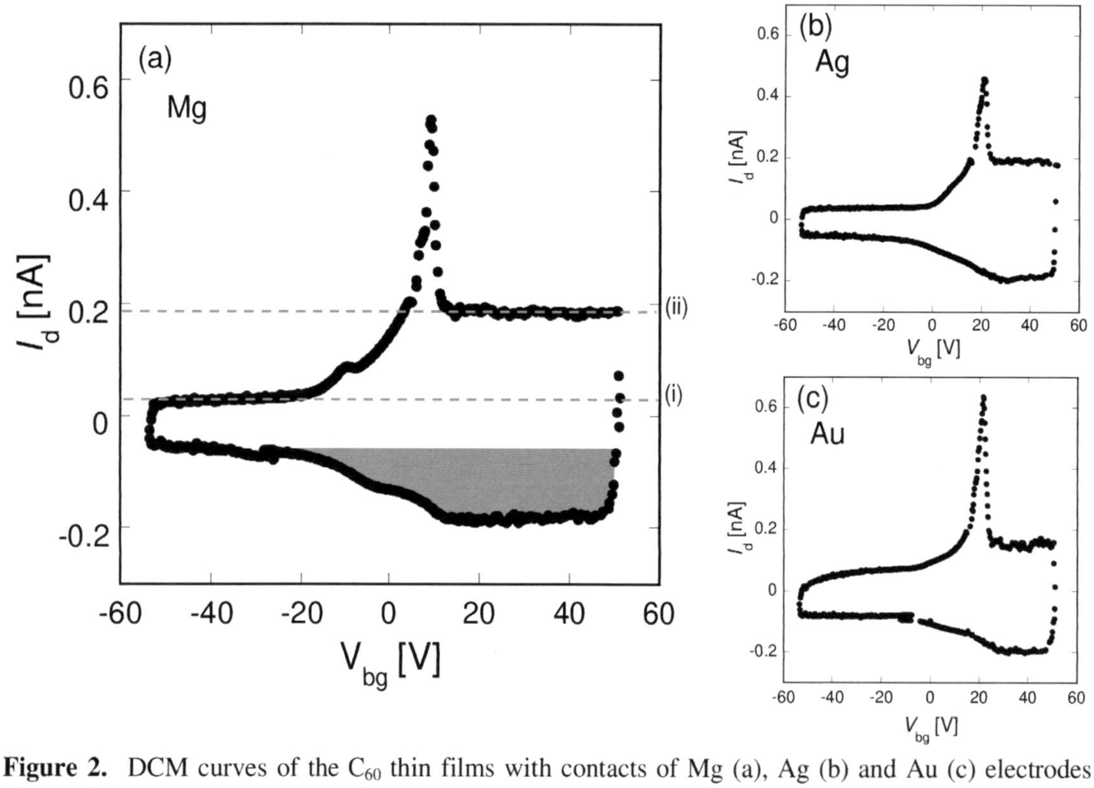

Figure 2. DCM curves of the C_{60} thin films with contacts of Mg (a), Ag (b) and Au (c) electrodes observed at room temperature. (i) and (ii) in Fig.2(a) show 0.021 nA and 1.8 nA, respectively. The area painted by gray is used for the estimation of the amount of injected carriers.

Table 1. FET and DCM characteristics of each sample with different contact metals.

Contact	μ	V_{th}	V_p	Injected electron [a]	Electron concentration [a]
Mg	0.62 cm^2/Vs	6 V	9.5 V	3.5×10^{-8} C	1.6×10^{12} $/cm^2$
Ag	0.34 cm^2/Vs	10 V	21.5 V	2.6×10^{-8} C	8.8×10^{11} $/cm^2$
Au	0.076 cm^2/Vs	11 V	22 V	2.0×10^{-8} C	6.9×10^{11} $/cm^2$

[a] These are estimated from integral of the displacement current curve and the area of the C_{60} thin film.

Another feature can be seen in the Mg sample that there is a shoulder before the peak in the DCM curve as shown in Fig. 2(a). A similar shoulder is observed also in the back sweep of V_{bg}. Such a clear two-stage transition can not be observed in the curves of the Ag and the Au sample at room temperature as shown in Figs. 2(b) and 2(c). In order to clarify the mechanism, the DCM curves of the Mg sample are measured at different temperatures as shown in Fig. 3. By decreasing temperature, the curves shift to the right (high V_{bg}) direction, consequently the shoulders of the lower transition also shift to the right. On the other hand, the I_d value at the shoulder decays with lowering temperature and the structure almost disappears at temperatures lower than 220 K. Comparing the metals, it is well known that alkali- and alkali-earth meals are easy to defuse into the C_{60} crystal and act as donor [7, 8]. Therefore, if we assume that Mg atoms have defused into the C_{60} layer and a doped layer would extend underneath the Mg electrode. In this scenario, Mg atoms form a certain impurity level under the LUMO level and then the electron can be donated from the Mg atoms with much smaller electric field at higher temperatures. However, at lower temperatures below 220 K, the thermal energy would not be sufficient to release electrons and only the injected electron from the electrode would be dominant to accumulate in the C_{60} layer. Therefore, such a two stage transition could not be observed at lower temperatures. However, further study is necessary to clarify the difference.

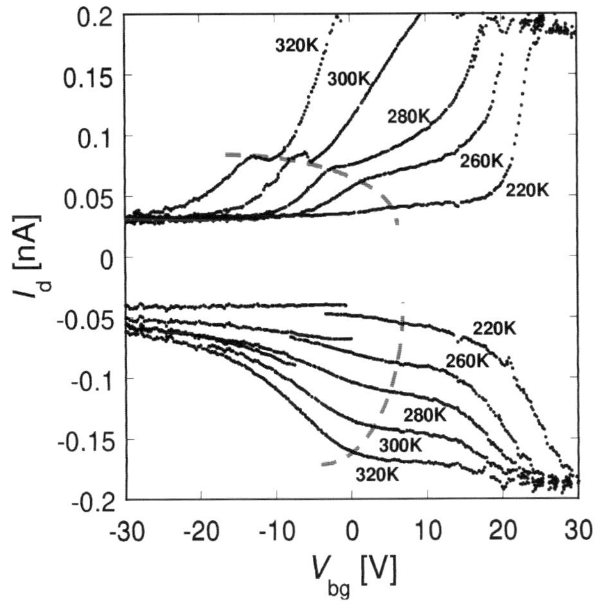

Figure 3. DCM curves of the Mg sample measured at different temperatures. The curves are magnified around the first transition. The broken lines are guide for eyes to understand the shift of the shoulders by decreasing of temperature.

Journal of Physics: Conference Series **159** (2009) 012014 doi:10.1088/1742-6596/159/1/012014

4. Summary

Carrier injection into a C_{60} thin film has been studied from the FET and the DCM characteristics in samples with contacts of Mg, Ag and Au. The Mg sample shows the highest mobility and the lowest V_{th} among the samples. A peak of displacement current is observed in the DCM curves of all the samples. The V_p shifts with the same trend of the V_{th} depending on the contact metals. Two-stage transition is observed only in the DCM curve of the Mg sample. It is confirmed that the DCM curve shifts to higher V_{bg} direction and the lower transition is smeared by decreasing the temperature. The transition would be caused by a diffusion of Mg atoms into the C_{60} layer and they could act as donor sites in the film.

Acknowledgments
The work was supported in part by Grants-in-Aid for Scientific Research from the Japan Society for the Promotion of Science (19054016, 16656007 and 16206001).

References

[1] Ishii H, Sugiyama K, Ito E and Seki K 1999 *Adv. Mater. (Weinheim, Ger.)* **11** 605
[2] Horiuchi K, Nakada K, Uchino S, Hashii S, Hashimoto A, Aoki N, Ochiai Y, Shimizu M 2002 *Appl. Phys. Lett.* **81** 1911
[3] Itaka K, Yamashiro M, Yamaguchi J, Haemori M, Yaginuma S, Matsumoto Y, Kondo M and Koinuma H 2006 *Adv. Mater. (Weinheim, Ger.)* **18** 1713
[4] Ochi K, Nagano T, Ohta T, Nouchi R, Kubozonoa Y, Matsuoka Y, Shikoh E and Fujiwara A 2006 *Appl. Phys. Lett.* **89** 083511
[5] Ogawa S, Naijo T, Kimura Y, Ishii H, and Niwano M 2005 *Synthetic Metals* **153** 253
[6] Lee S H, Yu Y S, Hwang S W and Ahn D 2007 *J. Nanosci. Nanotechnol.* **7** 4089
[7] Gunnarsson O 1997 *Rev. Mod. Phys.* **69** 575
[8] Brown C M, Taga S, Gogia B, Kordatos K, Margadonna S, Prassides K, Iwasa Y, Tanigaki K, Fitch A N and Pattison P 1999 *Phys. Lev. Lett.* **83** 2258

Molecular dynamics of fullerene-nanowhiskers studied by solid state NMR

H Ogata[1,2], S Motohashi[2] and S Tsuchida[2]

[1]Department of Chemical Science and Technology, Faculty of Bioscience and Applied Chemistry, Hosei University, 3-7-2, Kajino-cyou, Koganei, Tokyo, 184-8584, Japan

[2]Graduate School of Engineering, Hosei University, 3-7-2, Kajino-cyou, Koganei, Tokyo, 184-8584, Japan

E-mail: hogata@hosei.ac.jp

Abstract. The properties of molecular motion of as-grown C_{60} nanowhiskers and C_{70} nanowhiskers were investigated by ^{13}C-CP/MAS NMR, wideline ^{13}C-NMR and ^{1}H-NMR techniques. It was confirmed that *m*-xylene molecules are contained in both C_{60} and C_{70} nanowhisekers. It was found that C_{60} nanowhiskers exhibit an orientational order phase transition at 250 K. The activation energy of C_{60} molecules below 250 K was evaluated to be 14.7 kJ/mol, which is smaller than that of pristine C_{60} solid. For C_{70} nanowhiskers, no clear phase transition was found from 270 K to 170 K. The activation energy of C_{70} molecules was evaluated to be 13.7 kJ/mol between 270 K and 170 K, which is smaller than that of pristine C_{70} solid.

1. Introduction

Recently, new types of needle-like fullerene crystals called fullerene nanowhiskers were reported to be easily grown by the liquid-liquid interfacial precipitation method [1-9]. The typical C_{60} nanowhiskers are less than 0.5 μm in diameter and more than 100 μm in length. The preliminary results of C_{60} nanowhiskers by using a transmission electron microscope showed that the intermolecular distance along the growth axis is shorten by about 4%, compared with that of intrinsic bulk C_{60} crystals. For the crystal structure of C_{60} nanowhiskers, polymerized via the "2+2" cycloaddition model was proposed[2]. However, it is well known that C_{60} molecules in solids are chemically polymerized under irradiation with laser light or electron beam, or at high temperatures and pressures[10-14]. Tachibana et al reported that C_{60} molecules in the pristine nanowhiskers are bonded by van der Waals forces from Raman scattering measurements[4]. Minato *et al* reported the solvated structure of C_{60} nanowhiskers with hexagonal system[9]. For the crystal structure of C_{70} nanowhiskers, tetragonal structure was proposed[3,6]. However, the detailed structure and properties of molecular motion of both C_{60} nanowhiskers and C_{70} nanowhiskers are not fully understood yet. In this study, we report the properties of molecular motions of both C_{60} nanowhiskers and C_{70} nanowhiskers by means of solid-state ^{13}C-CP/MAS NMR, wideline ^{13}C- and ^{1}H-NMR measurements.

2. Experimental

C_{60} and C_{70} nanowhiskers used in this study were grown by a liquid-liquid interfacial precipitation method as previously reported[2-4]. The saturated solution of C_{60} or C_{70} and *m*-xylene was prepared and pored into a glass bottle. Then isopropyl alcohol was added into the bottle to form a liquid-liquid interface. The bottle was capped and kept in the refrigerator for 10 days under the illumination of the fluorescent light. The grown samples sank to the bottom of the bottle. The solvent of the as grown sample was dried by using a paper filter, and the sample was put in a glass tube or roter for NMR measurements and sealed immediately.

Solid-state ^{13}C-CP/MAS NMR spectrum was measured in a 7.05 T magnetic field using a AVANCE300WB spectrometer(Bruker) with zirconia roters(4 mm in diameter) spun in dried air at 9 kHz. Wide-line ^{13}C- and ^{1}H-NMR spectra and their spin-lattice relaxation times(T_1) by the saturation recovery technique were measured in a 9.4 T magnetic field using a DSX400 WB spectrometer (Bruker).

3. Results and Discussion

3.1. C_{60} nanowhiskers

Figure 1 shows the solid-state ^{13}C-CP/MAS NMR spectrum of as grown C_{60} nanowhiskers at 297 K. The peak at 143.8 ppm can be assigned to be the carbon on C_{60}. Peaks at 135.9, 131.2, 129.8, 126.6 and 23.4 ppm can be assigened to be the carbon on *m*-xylene. This fact clearly shows that *m*-xylene are contained in the C_{60} nanowhiskers. No peaks around 30 ppm, which are characteristic to the sp^3-like carbon[14-16], are observed. This fact suggests that there is no polymerized C_{60}-C_{60} intermolecular bonds in the C_{60} nanowhiskers as previously reported[2].

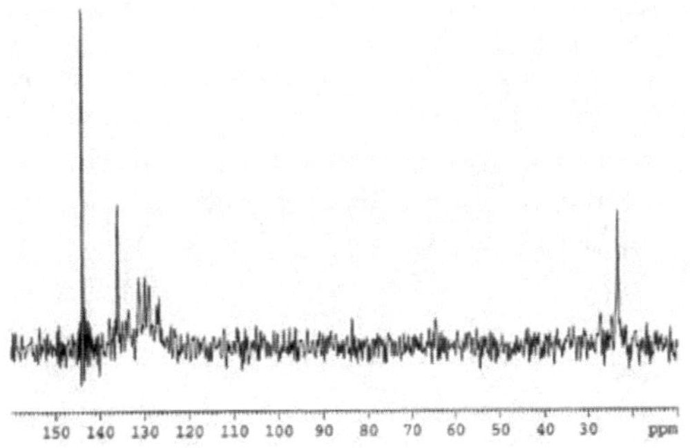

Figure 1. Solid-state ^{13}C-CP/MAS NMR spectrum of C_{60} nanowhiskers.

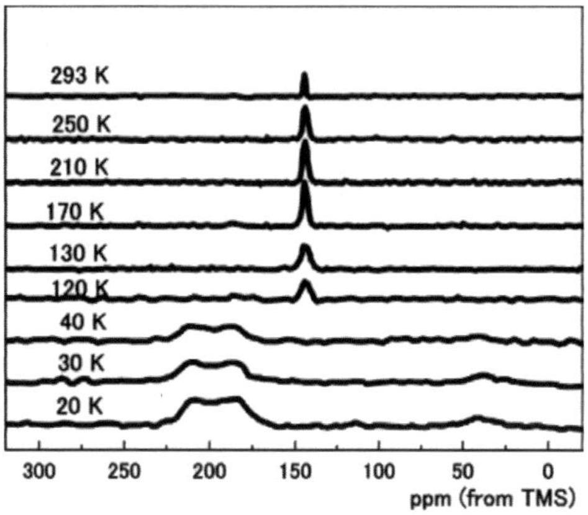

Figure 2. Wideline ^{13}C-NMR spectra of C_{60} nanowhiskers as a function of temperature.

Figure 2 shows wideline ^{13}C-NMR spectra of C_{60} nanowhiskers as a function of temperature. A sharp line observed near room temperature broadens slightly and exhibits asymmetry below 250 K. It is thought that this phenomenon is associated with the orient-ational order transition confirmed in the pristine C_{60} solid[17,18]. That is, the nature of the rotation of the C_{60} molecules in C_{60} nanowhiskers change from rapid isotropic reorientation to uniaxial jump reorientation at 250K. Below 120 K, the line gradually broadens and develops the structure characteristic of a chemical shift anisotropy powder pattern. The ^{13}C-NMR magnetization recovery data of C_{60} nanowhiskers could be fitted to a single exponential function above 250 K, on the other hand, to a biexponential function below 250 K as follows,

$$1 - M(t)/M(\infty) = F^S \exp\left(-t/T_1^S\right) + F^L \exp\left(-t/T_1^L\right) \quad (1)$$

with F^S, T_1^S, F^L, and T_1^L being the fitting parameters. Here, $M(t)$ is the magnetization at time t after saturation, $F^S(F^L)$ is the fraction of the shorter(longer) component of $T_1^S(T_1^L)$. In the case of C_{60} nanowhiskers, the dominant spin-lattice relaxation mechanism is considered to be provided by the chemical shift anisotropy (CSA), i.e., the fact that magnitude and direction of the magnetic field at a ^{13}C nucleus depend slightly on the orientation of the molecule as a result of anisotropic magnetic shielding by the electrons of that molecule[19]. Another possible spin-lattice relaxation mechanism is considered to be dipole interaction with ^1H nuclei of *m*-xylene molecules. We attribute the biexponential behavior below 250 K to the inhomogeneity of the dipole interaction with ^1H nuclei. Although the detailed information about the crystal structure is necessary to identify the relaxtion mechanism, it is plausible that uniaxial jump reorientation below 250 K causes the inhomogeneity of the dipole interaction with ^1H nuclei at carbon sites on C_{60} molecule. Assuming this model, T_1^L is associated with the CSA . The temperature dependence of T_1 (above 250 K), T_1^S and T_1^L (below 250 K) are shown in Figure 3.

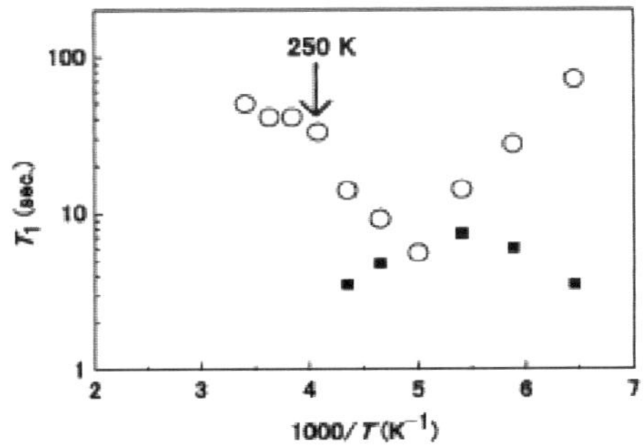

Figure 3. Temperature dependence of the ^{13}C- T_1 in C_{60} nanowhiskers (○:T_1^L, ■ : T_1^S).

From the temperature dependence of T_1^L, activation energy of C_{60} molecules below 250 K was evaluated to be 14.7 kJ/mol based on the CSA relaxation mechanism, which is smaller than that of pristine C_{60} solid reported to be 24.1 kJ/mol at simple cubic phase below 260 K[18]. This fact suggests that interaction between C_{60} molecules in C_{60} nanowhisker is smaller than that in pristine C_{60} solid due to *m*-xylene molecules in C_{60} nanowhisker crystal.

Figure 4(a) shows wideline ^1H-NMR spectra of C_{60} nanowhiskers as a function of temper-ature. Two peaks observed at 6.5 and 1.5 ppm at 293 K are attributed to the aromatic and methyl group proton in *m*-xylene molecule, respectively. As the temperature decreases, the line gradually broadens and slightly develops the triplet structure characteristic of a three spin system such as methyl proton.

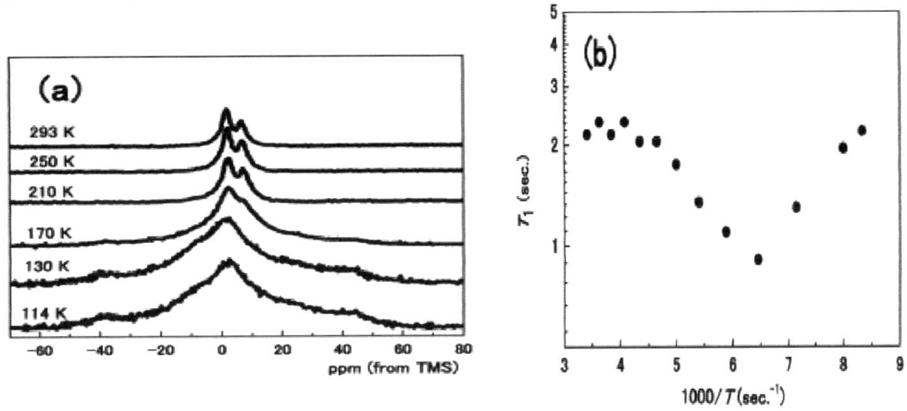

Figure 4. Temperature dependences of (a)^1H-NMR spectrum and (b)the ^1H-T_1 in C_{60} nanowhiskers.

The temperature dependence of ^{1}H-T_1 is shown in Figure 4(b). T_1 exhibits a minimum at 155 K of 0.9 sec. Assuming that the spin-lattice relaxation mechanism responsible for this minimum is caused by the modulation of dipole interaction caused by the methyl group reorientation around C_3 axis, then the activation energy for such motion was evaluated to be 7.0 kJ/mol from the expression of Bloembergen, Purcell, and Pound[20] as corrected by Kubo and Tomita[21]. This is an appropriate value as the thermal activated reorientation of methyl group around C_3 axis[22].

3.2. C_{70} nanowhiskers

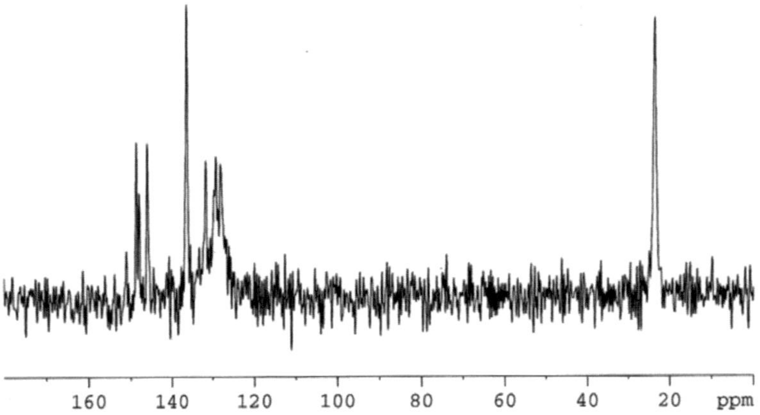

Figure 5. Solid-state ^{13}C-CP/MAS NMR spectrum of C_{70} nanowhiskers.

Figure 5 shows the solid-state ^{13}C-CP/MAS NMR spectrum of C_{70} nanowhiskers at 298 K. Peaks at 151.0,148.6,148.0,146.1 and 131.6ppm can be assigned to the carbon on C_{70}. Peaks at 136.3,129.2,128.1 and 23.4ppm are attributed to the carbon of the *m*-xylene. This fact clearly shows that *m*-xylene are contained in C_{70} nanowhiskers. No peaks around 30 ppm, which are characteristic to the sp^3-like carbon are also observed. This fact suggests that there is no polymerized C_{70}-C_{70} intermolecular bond in the C_{70} nanowhiskers.

Figure 6(a) shows the temperature dependence of wideline ^{13}C-NMR spectra of C_{70} nano-whiskers from 270K to 170K. No remarkable changes are observed in the lineshape. This behavior is quite different for the case of pristine C_{70} solid[23]. It is suggested that C_{70} nanowhiskers reveal no clear phase transition from 270 K to 170 K.

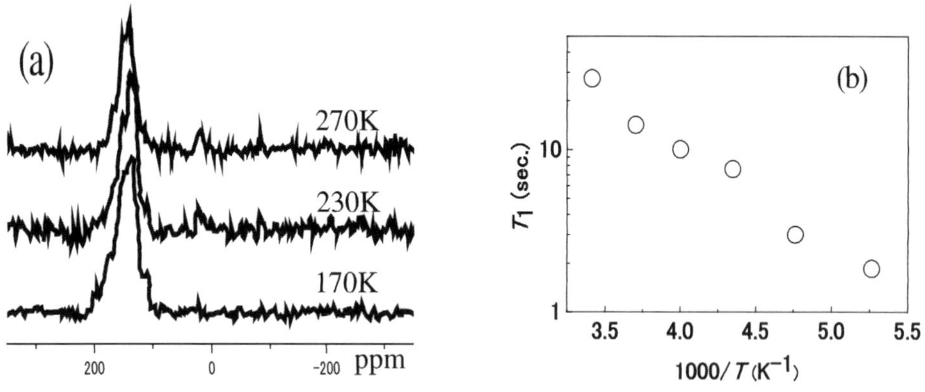

Figure 6. Temperature dependence of (a)^{13}C-NMR spectrum and (b)^{13}C-T_1 in C_{70} nanowhiskers.

The temperature dependence of ^{13}C-T_1 is shown in Figure 6(b). The activation energy of C_{70} molecular rotation between 270 K and 170 K was evaluated to be 13.7 kJ/mol based on the CSA relaxation mechanism, which is smaller than that of pristine C_{70} solid reported to be 22.0 kJ/mol below 280 K[23]. This fact suggests that intermolecular interaction between C_{70} molecules in C_{70} nanowhisker is smaller than that of pristine C_{70} solid.

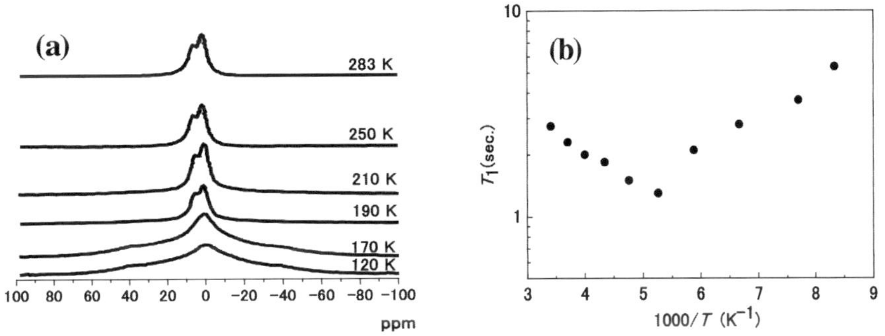

Figure 7. Temperature dependences of (a)^{1}H-NMR spectrum and (b)the ^{1}H-T_1 in C_{70} nanowhiskers.

Figure 7(a) shows wideline ^{1}H-NMR spectra of C_{70} nanowhiskers as a function of temperature. At 293 K, Two peaks observed at 6.4 and 1.6 ppm at 283 K are attributed to the aromatic and methyl group proton in *m*-xylene molecule, respectively. As the temperature decreases, the line gradually broadens and slightly develops the triplet structure characteristic of a three spin system such as methyl proton. The temperature dependence of ^{1}H-T_1 is shown in Figure 7(b). Assuming that the spin-lattice relaxation mechanism is caused by the modulation of dipole interaction caused by the methyl group reorientation around C_3 axis, then the activation energy for such motion was evaluated to be 6.7 kJ/mol from the expression of Bloembergen, Purcell, and Pound as corrected by Kubo and Tomita.

4. Conclusion

The properties of molecular motions of both C_{60} nanowhiskers and C_{70} nanowhiskers were investigated by means of solid-state NMR techniques. For C_{60} nanowhiskers, orientational order transition at 250 K was confirmed. The activation energy of C_{60} molecules below 250 K was evaluated to be 14.7 kJ/mol, which is smaller than that of pristine C_{60} solid. For C_{70} nanowhiskers, no clear phase transition was found from 270 K to 170 K. The activation energy of C_{70} molecules was evaluated to be 13.7 kJ/mol between 270 K and 170 K, which is smaller than that of pristine C_{70} solid. Inter-molecular interaction between fullerene molecules in nanowhisker is considered to be smaller than that of pristine fullerene solid due to the solvated *m*-xylene molecules.

References

[1] Miyazawa K, Obayashi A and Kuwabara M 2001 *J.Am.Ceram.Soc.* **84** 3037
[2] Miyazawa K, Kuwasaki Y, Obayashi A and Kuwabara M 2002 *J.Mater.Res.* **17** 83
[3] Miyazawa K 2002 *J.Am.Ceram.Soc.* **85** 1297
[4] Tachibana M, Kobayashi K, Uchida T, Kojima K, Tanimura M and Miyazawa K 2003 *Chem. Phys. Lett.* **374** 279
[5] Miyazawa K, Kuwasaki Y, Hamamoto K, Nagata S, Obayashi A and Kuwabara M 2003 *Surface and Interface Analysis* **35** 117
[6] Miyazawa K, Hamamoto K, Nagata S and Suga T 2003 *J.Mater.Res.* **18** 1096
[7] Miyazawa K, Mashino T and Suga T 2003 *J.Mater.Res.* **18** 2730

[8] Miyazawa K and Suga T 2004 *J.Mater.Res.* **19** 2410
[9] Minato J and Miyazawa K 2005 *Carbon* **43** 2837
[10] Rao A M, Zhou P, Wang K-A, Hager G T, Holden J M, Wang Y, Lee W-T, Bi X-X, Eklund P C, Cornett D S, Duncan M A and Amster I J 1993 *Science* **256** 955
[11] Matus M and Kuzmany H 1993 *Appl.Phys.*A **56** 241
[12] Matsuishi K, Ohno T, Yasuda N, Nakanishi T, Onari S and Arai T 1997 *J. Phys. Chem. Solids* **58** 1747
[13] Tachibana M, Sakuma H and Kojima K 1997 *J.Appl.Phys.* **82** 4253
[14] Iwasa Y, Arima T, Fleming R M, Siegrist T, Zhou O, Haddon R C, Rothberg L J, Lyons K B, Carter Jr H L, Hebard A F, Tycko R, Dabbagh G, Krajewski J J, Thomas G A and Yagi T 1994 *Science* **264** 1570
[15] Alloul H, Brouet V, Lafontaine E, Malier L and Forro L 1996 *Phys.Rev.Lett.***76** 2922
[16] Maniwa Y, Sato M, Kume K, Kozlov M E and Tokumoto M 1996 *Carbon* **34** 1287
[17] Heiney P A, Fischer J E, McGhie A R, Romanow W J, Denenstein A M, McCauley Jr J P, Smith III A B and Cox D E 1991 *Phys. Rev. Lett.* **66** 2911
[18] Tycko R, Dabbagh G, Fleming R M, Haddon R C, Makhija A V and Zahurak S M 1991 *Phys. Rev. Lett.* **67** 1886
[19] Abragam A 1961 *Principles of Nuclear Magnetism* (New York : Oxford Univ. Press) Chap.8
[20] Bloembergen N, Purcell E M and Pound R V 1948 *Phys.Rev.* **73** 679
[21] Kubo R and Tomita K 1954 *J.Phys.Soc.Japan* **9** 888
[22] Allen P S and Cowking A 1968 *J.Chem. Phys.* **49** 789
[23] Maniwa Y, Ohi A, Mizoguchi K, Kume K, Kikuchi K, Saito K, Ikemoto I, Suzuki S and Achiba Y 1993 *J.Phys. Soc.Jpn.* **62** 1131

Journal of Physics: Conference Series **159** (2009) 012016 doi:10.1088/1742-6596/159/1/012016

Self-assembling Fullerene Derivatives for Energy Transfer in Molecular Gel System

N Watanabe[1], H Jintoku[1], T Sagawa[2], M Takafuji[1], T Sawada[1] and H Ihara[1]

[1]Department of Applied Chemistry & Biochemistry, Kumamoto University, 2-39-1 Kurokami, Kumamoto 860-8555, Japan
[2]Instite of Advanced Energy, Kyoto University, Gokasho, Uji 611-0011, Japan

E-mail: ihara@kumamoto-u.ac.jp

Abstract. An L-glutamide-derived compound with a fullerene moiety was newly synthesized by the coupling of (6,6)-phenyl-C61 butyric acid with *N,N'*-didodecyl L-glutamide containing an amino alkyl group. In this paper, we describe that the solubility to an organic solvent was sufficiently improved by functionalization with L-glutamide and these fullerene derivatives could form ordered aggregates in organic solvents. It is also reported that the fullerene-containing derivatives were evaluated as a donor-acceptor model with the corresponding porphyrine derivative.

1. Introduction

Porphyrin and benzoquinone derivatives play as an electron donor-acceptor for light-harvesting and electron transfer in the natural photosynthesis. Recently, a porphyrin-fullerene complex has been of much concern as a mimic photosynthetic model, because the first reduction potential of fullerene is similar to that of benzoquinone and fullerene is expected to be a potential electron acceptor for a π-electron system [1]. For example, it is known that the lifetime of a charge-separated state in a porphyrin-fullerene conjugate is longer than that in the natural photosynthetic system [2].

In this paper, we wish to introduce a new porphyrin-fullerene complex system that is tuned by functionalization with L-glutamide derivatization. It has been reported that an L-glutamide-derived moiety is a very useful tool for promoting the intermolecular hydrogen bonding to lead to suprastructural fabrications [3]. Therefore, we newly synthesized three kinds of L-glutamide-functionalized fullerenes **1** (**FC2**, **FC5** and **FC10**) (figure 1) and then they were evaluated as a donor-acceptor model with the corresponding porphyrin derivative **2** [4].

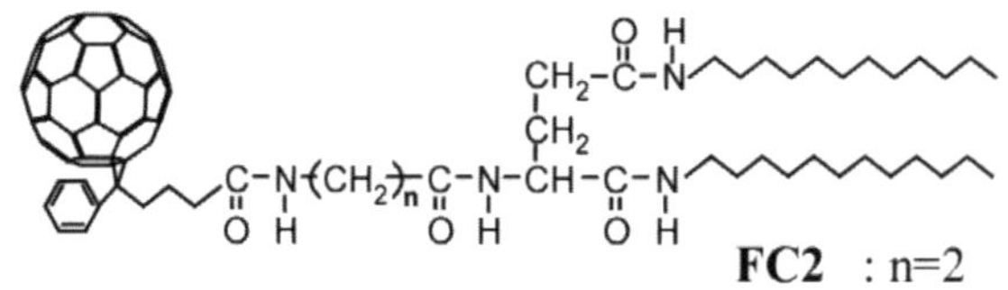

FC2 : n=2
FC5 : n=5
FC10 : n=10

1

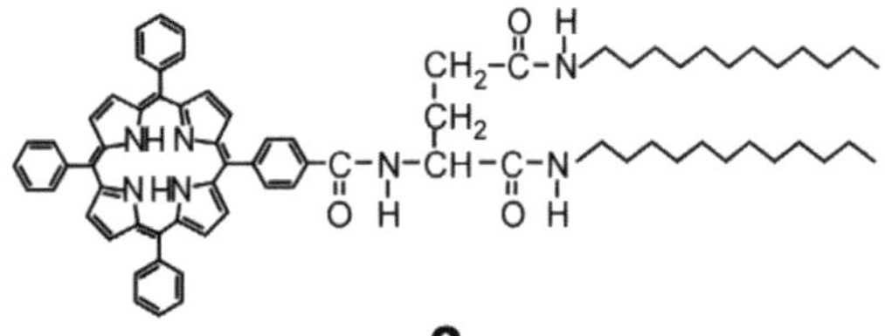

2

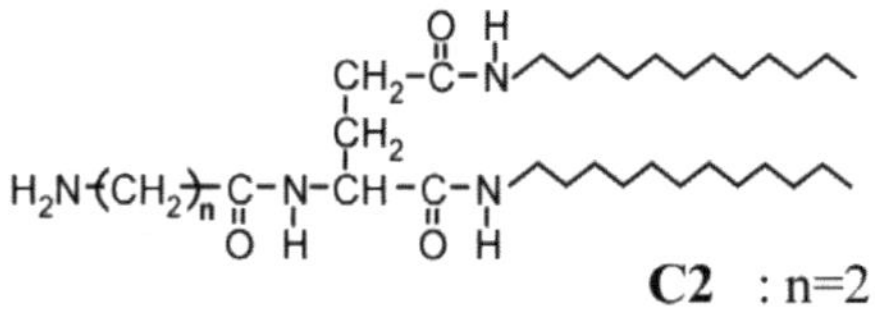

C2 : n=2
C5 : n=5
C10 : n=10

3

Figure 1. L-Glutamide-derived lipids containing fullerene (**1**), porphyrin (**2**) and alkyl amino group (**3**).

2. Experimental

N,N'-didodecyl L-glutamide containing an amino alkyl group **3** (**C2**, **C5** or **C10**) was synthesized according to the previously reported method [5].

(6,6)-phenyl-C61 butyric acid (PCBA) was prepared by (6,6)-phenyl-C61 butyric acid methyl ester (PCBM) [6]. Acetic acid and HCl were added to a solution of PCBM in toluene. The mixture was stirred and refluxed. The time course of reaction was followed by TLC. All volatile components were removed in *vacuo*. The residue was dispersed by methanol and then washed by centrifugation with ether. The same procedure was repeated by using toluene and ether. The residue was dried in *vacuo* to give PCBA as dark brown powders. The chemical structures were confirmed by elemental analysis, FT-IR, UV-vis, MALDI-TOF mass and HPLC. PCBA was dissolved in pyridine and **3** (**C2**, **C5** or **C10**) and *N,N'*-dicyclohexylcarbodiimide (DCC) were added to the PCBA solution. The mixture was stirred and refluxed. The time course of reaction was followed by HPLC. The reaction mixture was concentrated in *vacuo*. The residue was dissolved in dichloromethane and was filtered. The filtrate was concentrated in *vacuo*, and the residue was reprecipitated from toluene-acetonitrile and then dried in *vacuo* to give **1** (**FC2**, **FC5** or **FC10**) as brown powders. The chemical structures were confirmed by elemental analysis, ^{1}H and ^{13}C NMR, FT-IR, UV-vis, MALDI-TOF mass and HPLC.

3. Result and Discussion

3.1. Solubility and gelation property

$\mathbf{C_{60}}$ shows limited solubility. As shown in table 1 and figure 2, the solubility of $\mathbf{C_{60}}$ is considerably modified by derivatization into PCBM and introduction of an L-glutamide moiety. Especially the L-glutamide-functionalized $\mathbf{C_{60}}$, **1** showed good solubility for THF, toluene and chloroform. However, when a solvent system was adjusted by mixing with a poor solvent such as cyclohexane, the solution of **1** changed to a gel state. Figure 2c shows that **1** (**FC10**) undergoes a temperature-dependent phase transition between gel and sol states in cyclohexane-THF (6 : 1) mixture. This is a typical phenomenon in a molecular gel induced by L-glutamide-derived lipids [3]. We have reported that various L-glutamide derivatives form gels through nanofibrillar network formation. As supporting this, the TEM observation indicated that **1** could form nanostructural aggregates.

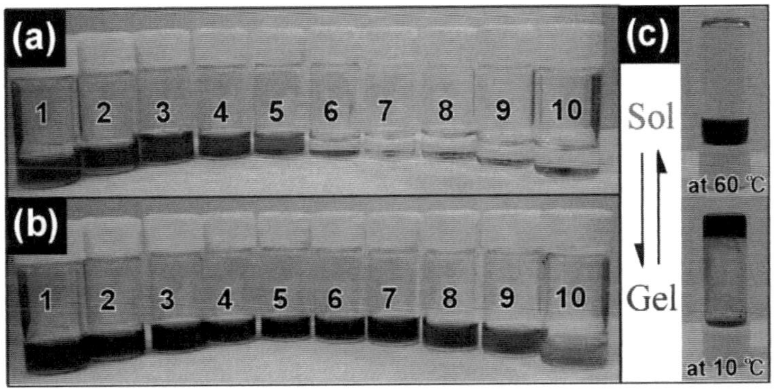

Figure 2. Solubility of C_{60} (a) and **FC5** (b) at 1.0 mM in 1: benzene, 2: toluene, 3: dichlorobenzene, 4: dichloroethane, 5: chloroform, 6: THF, 7: 1,4-dioxane, 8: 1-propanol, 9: 2-propanol and 10: ether at 25 °C. Temperature-dependent sol-to-gel transition of **FC10** (5.0 mM) in a cyclohexane-THF (6 : 1) mixture (c).

Table 1. Solubility of fullerene derivatives

solvent	C_{60}	PCBM	PCBA	FC5
toluene	S+	S+	I	S++
benzene	S+	S+	I	S+
dichlorobenzene	S+	S++	I	S++
dichloroethane	S	S++	I	S++
chloroform	S	S+	I	S++
tetrahydrofuran	I	S	S	S++
1,4-dioxane	I	S	I	S+
1-propanol	I	S	I	S
2-propanol	I	S	I	S
ether	I	I	I	S
water	I	I	I	I
methanol	I	I	I	I
acetonitrile	I	I	I	I
cyclohexane	I	I	I	I

S: soluble, I: insoluble
Concentration: 1.0 mM, Temperature: 25°C

3.2. Photophysical property

We have already reported that **2** can form a gel in organic media [4]. The typical phenomena were detected by UV-vis and CD spectral changes. For example, the Soret band of **2** appeared at 417 nm as a peak top in a solution state, but changed to a characteristic split by about 30 nm (398 nm and 428 nm) as peak tops in a gel state. In addition, the gelation was accompanied by induction of strong Cotton effect in 375 - 460 nm.

In this study, a photophysical property was investigated in a mixed system of **1** and **2**. After a cyclohexane-THF (20 : 1) mixture containing **1** and **2** was kept at 60 °C for 10 min and cooled down at 10 °C, UV-vis and CD spectra were examined. A similar procedure was carried out with a **2** and C_{60} mixed system. As shown in figures 3a and b, the **2** and C_{60} (2 : 1) system showed remarkable decrease of porphyrin absorption with weak Cotton effect. This is due to precipitation because of less solubility. In contrast, the **2** and **1** (**FC2**) (2 : 1) mixed system showed no precipitation but also typical splitting in UV-vis spectra with induction of strong Cotton effects at the absorption band around a porphyrin moiety as shown in figures 3c and d. These spectral changes were similar to that in **2** alone, but new Cotton effects appeared around 230 - 300 nm and 500 - 560 nm in the **2** and **FC2** mixed system (figure 3d).

On the other hand, addition of **1** (**FC2**) to the **2** solution (0.01 mM) in acetonitrile-benzene (100 : 1) induced a bathochromic shift of the Soret band (figure 4a). It is also observed that this bathochromic behavior was temperature-dependent (figure 4b). These results indicate that the **2** and **FC2** mixed system forms cooperative assemblies in mixed solvents such as cyclohexane-THF and acetonitrile-benzene at 10 °C.

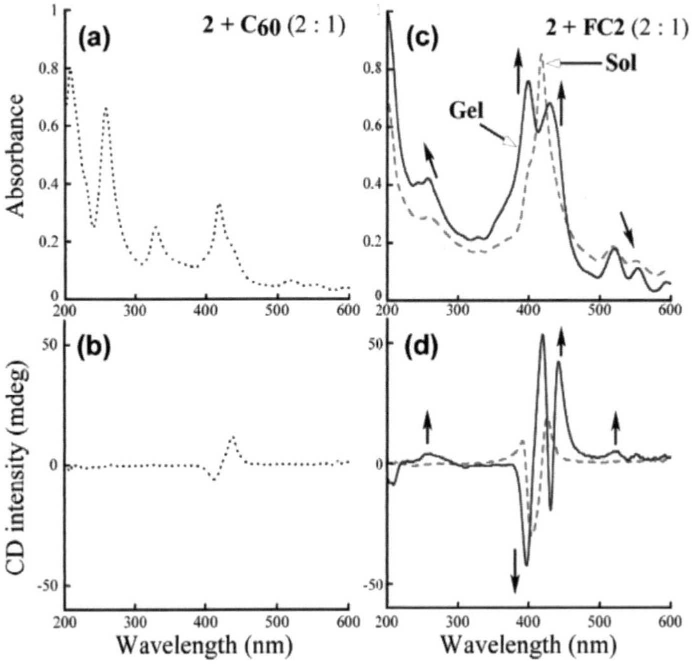

Figure 3. UV-vis (a, c) and CD (b, d) spectra of a **2** and C_{60} (2 : 1) mixture (dotted line) and a **2** and **FC2** (2 : 1) mixture (sol state: broken lines, gel state: solid line) in a cyclohexane-THF (20 : 1) mixture at 10 °C. Concentration **2**: 0.5 mM.

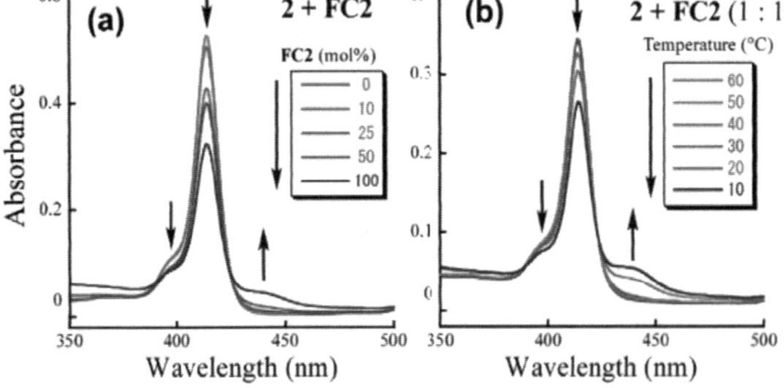

Figure 4. Concentration-dependent UV-vis spectral change of **2** (0.01 mM) in acetonitrile-benzene (100 : 1) at 25 °C (a). Temperature-dependent UV-vis spectral change of a **2** and **FC2** (1 : 1) mixture in acetonitrile-benzene (100 : 1) (b). Concentration **2**: 0.01 mM.

3.3. Energy transfer system

Well-defined electron donor-acceptor arrays in the natural photosynthetic reaction center ensure highly efficient photo-induced electron transfer (PET). Therefore, we investigated fluorescence characteristic of **2** with **1** (**FC2**) or $\mathbf{C_{60}}$ in acetonitrile-benzene (100 : 1). Figure 5a shows the concentration-dependent fluorescence spectral change at 25 °C. The emission intensity of **2** decreased remarkably with increase of **FC2**. Similar results are observed in the presence of $\mathbf{C_{60}}$ and no significant difference is seen in both the systems. This indicates the possibility of electron transfer from **2** to **FC2** and $\mathbf{C_{60}}$. On the other hand, the distinct difference between the **2** and $\mathbf{C_{60}}$ mixed systems was found in their temperature dependencies. As shown in figures 5b – 5d, the **2** and **FC2** (1 : 1) system showed remarkable temperature dependence showing the bending point at 40°C although almost no temperature dependence was observed in the $\mathbf{C_{60}}$ mixed system at the temperature range of 10 – 60 °C. This suggests that efficiency of the electron transfer can be promoted by temperature in the mixed system with **FC2**. To understand these specific behaviors, we propose that co-aggregation with ordered structures was promoted by mixing with **FC2** having an L-glutamide moiety. As supporting this assumption, we have reported that L-glutamide-derived lipids with hydrophobic functional head groups can form aggregates in organic media [7], showing a temperature-induced phase transition between ordered and disordered states.

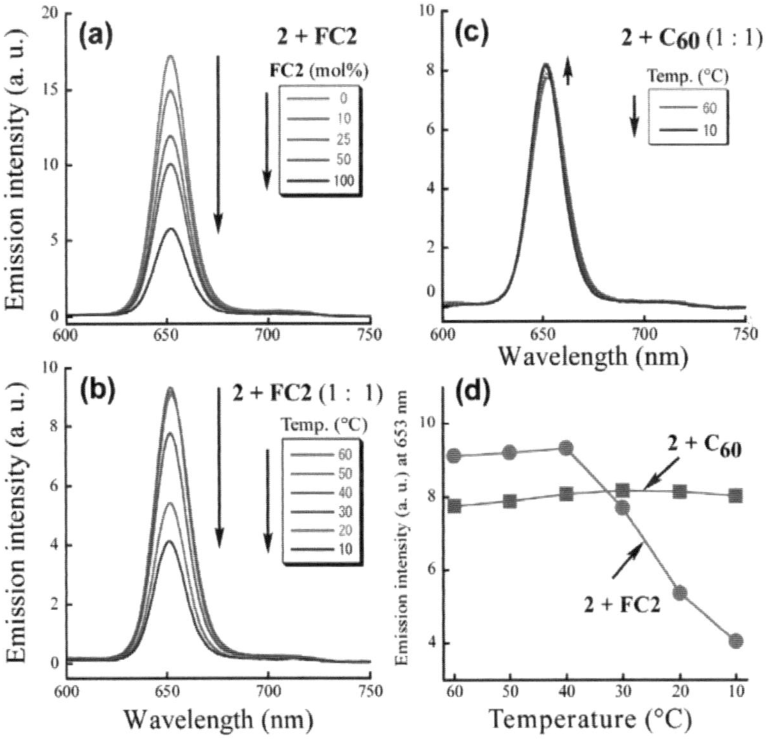

Figure 5. Concentration-dependent fluorescence spectral change of emission spectra of **2** (0.01 mM) in acetonitrile-benzene (100 : 1) at 25 °C (a). Temperature-dependent fluorescence spectral change of a **2** and **FC2** (1 : 1) mixture (b) and a **2** and $\mathbf{C_{60}}$ (1 : 1) mixture (c). Concentration **2**: 0.01 mM. Excitation wavelength = 435 nm. Temperature dependence of emission intensity at 653 nm (d).

4. Conclusion

It has been described that the L-glutamide-functionalized fullerene derivative (**1**) can dissolve in organic media to form a molecular gel through molecular orientation. A mixed system of the fullerene (**1**) and porphyrin (**2**) derivatives was applied for an electron transfer model and it was indicated that porphyrin-fullerene complexation could be assisted by an L-glutamide moiety.

Acknowledgments

This work was supported in part by a Grant-in-Aid for Scientific Research from the Ministry of Education, Science, Sports, and Culture of Japan.

References

[1] Umeyama T and Imahori H 2006 *Photosynthesis Res.* **87** 63
[2] Guldi D, Imahori H, Tamaki K, Kashiwagi Y, Yamada H, Sakata Y and Fukuzumi S 2004 *J. Phys. Chem.* **108** 541
[3] Takafuji M, Ishiodori A, Yamada T, Sakurai T and Ihara H 2004 *Chem. Commun.* **9** 1122; Ihara H, Takafuji M and Sakurai T 2004 *Enc. Nanosci. Nanotech.* **9** 473
[4] Jintoku H, Sagawa T, Sawada T, Takafuji M, Hachisako H and Ihara H 2008 *Tetrahedron Lett.* **49** 3987
[5] Ihara H, Hachisako H, Hirayama C and Yamada K 1987 *Liq. Crystals* **2** 215
[6] Hummelen J, Knight B, LePeg F and Wudl F 1995 *J. Org. Chem.* **60** 532
[7] (a) Yamada T, Derakhshan M, Ansarian R, Takafuji M, Hachisako H, Sagawa T and Ihara H 2005 *J. Mat. Res.* **20** 2486; (b) Hachisako H, Ihara H, Kamiya T, Hirayama C and Yamada K 1997 *Chem. Commun.* **1** 19; (c) Ihara H, Sakurai T, Yamada T, Hashimoto T, Takafuji M, Sagawa T and Hachisako H 2002 *Langmuir* **18** 7120

Electronic transport properties of photo-irradiated C_{60}

Y Chiba[1], S -R Chen[1, 2], H Tsuji[1], M Ueno[1], N Aoki[1] and Y Ochiai[1]

[1]Graduate School of Advanced Integration Science, Chiba University, Chiba, Japan
[2]Southern Taiwan University, Tainan, Taiwan

E-mail: ochiai@faculty.chiba-u.jp

Abstract. We have investigated field-effect transistor (FET) application on the photo-irradiated C_{60} thin film. Carrier mobility can be estimated to be 3.5×10^{-4} cm^2/Vs after photo-irradiation and it shows a lower mobility than that before irradiation. Probably, it would be due to introduction of many cracks in the C_{60} thin film induced by polymerization due to ultraviolet light irradiation. However, considering the decrease in the threshold voltage from 12 to 2 V in the I_{SD}-V_G characteristics, carrier injection barrier clearly decreases by photo-irradiated polymerization. Furthermore, the photo-irradiated C_{60} FET works even in atmosphere.

1. Introduction

Recently, one of the attractive topics in fullerene research is polymerization of C_{60} molecules. Since the first report in 1993 [1], several kinds of polymerization methods have been proposed [2-4]. Especially, photo- and electron-beam irradiations have advantageous applicabilities for nano-fabrications in the easiness and the area selectable. It is very important that those methods would have a possibility to realize the C_{60}-based carbon nano-devices. However, the electronic transport properties of polymerized C_{60} have not been yet investigated sufficiently. In this paper, therefore, we have discussed photo irradiation effect on the electronic transport properties of C_{60} thin film. It can be clarified by investigating the field-effect transistor (FET) characteristics. Furthermore, since we must determine the degree of the irradiation effect, the morphology of photo-irradiated C_{60} thin film has been studied by scanning electron microscopy (SEM) inspection.

2. Sample and experimental setup

Usually, in the transport experiment, a top-contact structure shows a good lead-connection in an organic-FET compared with a bottom-contact lead one. However, in our previous research, the electrodes for the top-contact blocked off the UV-light and polymerization is obstructed underneath the electrods. Therefore, in this study, the C_{60} FETs were fabricated with a bottom-contact configuration on p^{++}-doped Si substrates. The Si substrate had a 600 nm thick layer of thermally grown SiO_2 that was used for the gate insulator, and the substrate was used for a back gate electrode. Firstly, Ti/Au electrodes for source and drain were deposited 20 nm/80 nm, respectively, on the SiO_2 layer by electron-beam evaporation. The channel length L and the channel width W were 5.0 µm and 1.0 mm, respectively. Secondly, a C_{60} thin film of 100 nm thick was deposited on these electrodes by thermal evapolation with the substrate being kept at 440 K. Subsequently, photo-irradiation was performed for approximately 10 hours using the UV-visible light which was used for photolithography with a condition over 100 mW/cm^2, and at vacuum of ~10^6 Torr. The temperature of the sample was kept at 300 K during exposure in order to prevent a heating effect by photo-irradiation, because polymerized-C_{60} molecules by photo-irradiation could return to C_{60} monomers by heat treatment. The polymerization was checked with the degree of solubility in a toluene, as reported that such a polymerized C_{60} becomes insoluble in organic solvent [5]. The electrical measurements were performed at room temperature in vacuum (~10^6 Torr).

3. Results and discussion

Source-drain current (I_{SD}) versus source-drain voltage (V_{SD}) characteristics at several back-gate voltages (V_G) are shown in Fig. 1(a) for before the irradiation and Fig. 1(b) for after the irradiation. The field-effect mobility (μ) before photo-irradiation was 1.1×10^{-3} cm^2/Vs estimated from the fitting of the linear region of the I_{SD}-V_{SD} curves. We consider that this lower μ value than that in common studies on C_{60} FET mainly stemed from its bottom contact configration. We reffer to that, in our previous study on C_{60} FET with top contact configration, mobilites attained up to 0.6 cm^2/Vs [6]. The field-effect mobility after the irradiation was estimated to be 3.5×10^{-4} cm^2/Vs. In the results of Figs. 1(a) and (b), it has been confirmed that the I_{SD} and μ usually decreased by one order of magnitude after the irradiation. It has been reported that the mobility of photo-induced C_{60} polymer decreases because delocalized sp^2 bond transforms localized sp^3 bond [5]. Therefore, it is possible that the drops in the output-characteristics induced by photo-irradiation originated in this bond-transformation. However, for this reseach, it seems that the decreases of the I_{SD} and μ would be brought by the morphological modification in C_{60} thin film.

Subsequently, the I_{SD}-V_G characterisitcs at V_{SD} = 10V are shown in Fig. 1(c) for before (filled triangle) and after the irradiation (filled circle). The threshold-voltage (V_T) determined from I_{SD}-V_G curve was lowered from 12 to 2 V by the irradiation. It indicates that the injection barrier between the Au electrode and the C_{60} thin film had been changed by photo-polymerization, and consequently it must act in a good direction to make easy carrier injection. This result is qualitatively identical to the previous report that the band gap of photo-induced C_{60} polymer became smaller by one third than that of C_{60} solid [7]. This V_T of 2 V is considerably low compared with common C_{60} FET devices.

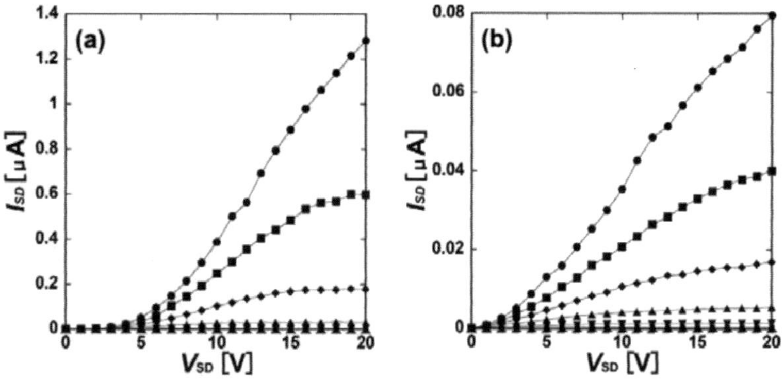

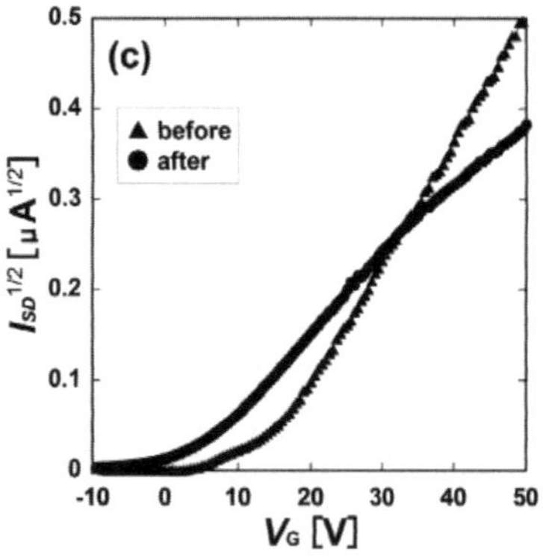

Figure 1. The FET characteristics of the C_{60}-thin film before and after photo-irradiation. I_{SD}-V_{SD} characteristics for before (a) and after the irradiation (b) are shown. The V_G was varied from 0 to 30 V. The mobilities decreased by one order of magnitude after the irradiation. I_{SD}-V_G characteristics for before (filled triangle) and after (filled circle) the irradiations are shown (c). The V_T was clearly lowered from 12 to 2 V by the irradiation.

Interestingly, it has been confirmed that the photo-irradiated C_{60} thin-film shows a clear FET characteristics even in the ambient atmosphere without any passivations [6] as shown in Fig. 2. Also, the FET working has been tested in the atomosphere. Although, the I_{SD} value was a lagery decreased three or four orders of magnitude, it still exhibited an obvious response in the back gate operation. The mobility was estimated to be 6.2×10^{-8} cm^2/Vs. On the photo-polymerized C_{60}, it has been reported that electrical transport in the ambient atmosphere has been observed in Ref. 8 but not observed in Ref. 5. Therefore, further investigations are needed for a more detailed understanding of this issue.

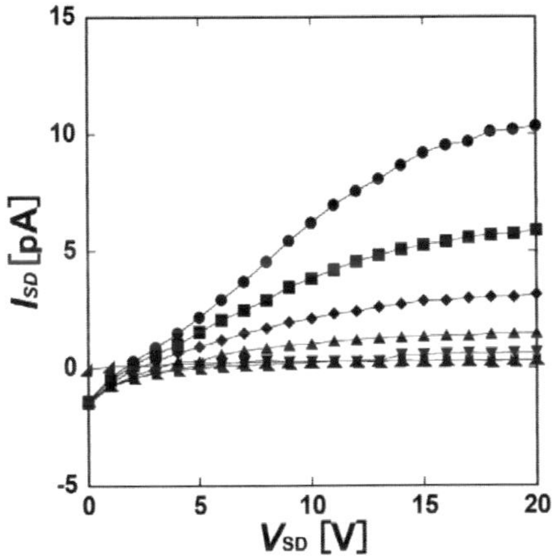

Figure 2. I_{SD}-V_{SD} characteristics of photo-irradiated C_{60} thin-film FET after exposure to the atmosphere is shown here. The V_G is varied from 0 to 30 V. Clear dependances on applied gate voltage are observed.

So as to determine the origin of a large decrease of the mobility, we have tried to check on the surface of the thin-film after photo-irradiation. In order to confirm the surface morphology after the irradiation, the surface of our C_{60} thin-film was investigated by using SEM. After the irradiation, there exsist many small cracks in the surface as shown in Fig. 3 and they may divide the C_{60} thin film into many domain areas with the scale of several micrometers on the entire film. Since these cracks could not be observed in a pristine C_{60} thin film, they should be caused by the irradiation. It has been reported that the intermolecular distance between C_{60} molecules is shortened about 0.1 nm by the polymerization [9]. The cracks would be caused by such shrinkage due to the polymerization, and consequently the mechanical disconnection could lead to a decrease in I_{SD} and μ in the FET. It can expect that the decrease of the μ should be caused by a large carrier scattering in the domain walls. Therefore the intrinsic transport may not be so largely affected by photo-irradiations. Here, in order to clarify such an intrinsic electrical transport in each small domain, we must have further many efforts for the determination of intra segment transport using another skillful technique.

Figure 3. The SEM photographs of C_{60} thin film after photo-irradiation. The inset is a magnified image of the main photograph. Many cracks can be observed over the film. It seems that these cracks are due to the decreases of intermolecule distance between neighboring C_{60} molecules induced by photo-polymerization.

4. Summary

The FET characteristics of C_{60} thin film were investigated before and after photo-irradiation. The μ decreases from 1.1×10^{-3} to 3.5×10^{-4} cm^2/Vs after the photo-irradiation. We consider that the decrease may be due to many small-domain-defects observed in SEM image on the photo-irradiated C_{60} thin film. These cracks must be due to the shrinkage of inter-molecule distance by polymerization. It indicates that carrier scatterings at the domain boundaries connect into the low μ value. It suggests that the μ is strongly affected with these cracks. Therefore, in order to clarify the intrinsic transport characteristics of photo-irradiated C_{60}, further precise investigation should be required on the polymerization. As for V_T, photo-irradiation makes a clear lowering on V_T from 12 to 2 V. It indicates that there exists the band gap modification of C_{60} due to the photo-polymerization and an improved carrier injection from Au electrodes to C_{60} active layer. It is confirmed that the photo-irradiated C_{60} thin-film exhibits obvious FET performance even in the ambient atmosphere. Although, the μ was considerably low after exposure to atmosphere, it may improve with optimized condition. Finally, this low value of V_T may encourage regarding a realization of low driving voltage C_{60} devices. Nevertheless, photo-polymerized C_{60} should overcome the fatal problem of *n*-type organic semiconductors in the atmosphere.

Acknowledgments

The work was supported in part by Grants-in-Aid for Scientific Research from the Japan Society for the Promotion of Science (19054016, 16656007 and 16206001).

References

[1] Rao A M, Zhou P, Wang K -A, Hager G T, Holden J M, Wang Y, Lee W –T, Bi X –X, Eklund P C, Cornett D S, Duncan D S and Amster I J 1993 *Science* **259** 955
[2] Yamawaki H, Yoshida M, Kakudate Y, Usuda S, Yokoi H, Fujiwara S, Aoki K, Ruoff R, Malhotra R and Lorents D 1993 *J. Phys. Chem.* **97** 11161
[3] Chauvet O, Oszlànyi G, Forro L, Stephens P W, Tegze M, Faigel G and Jànossy A 1994 *Phys. Rev. Lett.* **72** 2721
[4] Onoe J, Nakayama T, Aono M and Hara T 2003 *Appl. Phys. Lett.* **82** 595
[5] Dzwilewski A, Wågberg T and Edman L 2007 *Phys. Rev. B* **75** 075203
[6] Horiuchi K, Nakada S, Uchino S, Hashii S, Hashimoto A, Aoki N, Ochiai Y and Shimizu M 2002 *Appl. Phys. Lett.* **81** 1911
[7] Okada S and Saito S 1999 *Phys. Rev. B* **59** 1930
[8] Onoe J, Nakayama T, Aono M and Hara T 2004 *J. Phys. Chem. Sol.* **65** 343
[9] Bashkin I O, Rashchunpkin V I, Gurov A F, Moravsky A P, Rybchenko O G, Kobelev N P, Soifer Y M and Ponyatovski E G 1994 *J. Phys.: Condens. Matter* **6** 7491

Journal of Physics: Conference Series **159** (2009) 012018 doi:10.1088/1742-6596/159/1/012018

Structural Characterization of Mg-doped C_{60} Thin Films

N Kojima, M Natori, H Suzuki and M Yamaguchi

Toyota Technological Institute, 2-12-1 Hisakata Tenpaku, Nagoya 468-8511, Japan

E-mail: nkojima@toyota-ti.ac.jp

Abstract. The epitaxial growth of Mg-doped C_{60} films has been investigated. It is found that the epitaxial growth of Mg-doped C_{60} film is enabled by using mica (001) substrate in the low Mg concentration region (Mg/C_{60} molar ratio $<$ 1). The crystal quality of the epitaxial Mg-doped C_{60} film is improved drastically in compared with micro-crystalline film on glass substrate. Such drastic improvement of crystal quality in the epitaxial films results significant increase in conductivity. This result may indicate the significant increase of carrier mobility. In addition, the crystal structure change is observed in high Mg concentration region (Mg/C_{60} molar ratio $>$ 2). The C_{60} polymer sheet formation is suggested.

1. Introduction

C_{60} solids have been known as high resistive semiconductor materials. For organic solar cell applications, such high resistivity is one of the reasons of low charge transport efficiency in a C_{60} accepter layer. Impurity doping in organic semiconductor can increase conductivity, and control band profile at the interface between organic/metal or organic/organic materials. Previously, metal-doped C_{60} has been studied extensively with regard to the superconducting compounds. Alkali and alkaline earth metals, such as Li, Na, K, Rb and Ba have been used mostly as guest metals in the previous study. The valence electron level of most alkali and alkaline earth metals lies at higher energy than LUMO-derived band of C_{60}, and the valence electrons of guest metals transfer to the C_{60} LUMO-derived band. Thus, these compounds show the metallic behavior. For the semiconductor device application, the search for new guest materials is necessary. Mg is one of the promising materials for guest metal showing semiconductor property. R. P. Gupta *et al.* reported the energy band calculation of Mg_2C_{60} solids, and indicated that Mg_2C_{60} was semiconductor, since Mg 3s-derived occupied band was formed between C_{60} HOMO-LOMO levels [1]. In experimentally, it was reported that the conversion efficiency of C_{60}/MEH-PPV organic cells was significantly improved by the automatic Mg-doping in a C_{60} layer during the Mg electrode deposition [2, 3]. However, the detailed discussion of semiconductor properties of Mg-doped C_{60} films has not been reported yet.

The electric conduction of organic materials is also affected by the crystal quality. The band conduction mechanism has been observed in organic single-crystal materials, while the amorphous one shows the hopping conduction mechanism. Therefore, to investigate the essential electric conductivity behaviors, the well-crystalline film is required.

In this paper, we investigate the epitaxial growth of Mg-doped C_{60} films to obtain the well-crystalline films, and electric properties of the epitaxial Mg-doped C_{60} films.

2. Experimental

C_{60} and Mg-doped C_{60} films were grown on mica (001) or quarts glass substrates by molecular beam epitaxy (MBE). The base pressure of the MBE chamber was $3x10^{-7}$ Pa. The pure (99.98%) C_{60} powder and Mg (99.9%) were evaporated from a Knudsen cell. The beam flux of the each source was monitored by a nude ion gauge at the substrate position, and was controlled by the source temperature. The C_{60} beam flux was fixed at around $6.7x10^{-6}$ Pa. The film composition of Mg/C_{60} was changed by the changing the Mg beam flux from 0 to $2.7x10^{-6}$ Pa. The resulted film composition of Mg/C_{60}, estimated by X-ray Photoelectron Spectroscopy (XPS), was proportional to the beam flux ratio of Mg/C_{60}. The growth rate of Mg-doped C_{60} was around 5 nm/min, and the total film thickness was around 1 μm.

In our previous work, we have already established the growth conditions for epitaxial growth of pure C_{60} films on mica (001) substrate. (111)-oriented crystalline C_{60} can be grown on mica (001) at the growth temperature of 165°C. Therefore, we selected the same growth temperature for achieving the epitaxial growth of Mg-doped C_{60} films. Figure 1 shows the schematic model of molecular arrangement of C_{60} on mica (001) surface.

For the conductivity measurements, bottom electrode structure was used. MgAg or Al metal was evaporated through a metal mask on the substrate to form a pair of electrodes with 0.5 mm spacing, before Mg-doped C_{60} was grown. The conductivity was estimated from I-V characteristics between the adjacent bottom electrodes under the vacuum condition.

The crystal structure was analyzed by X-ray diffraction (XRD) measurements with two configurations. One is 2θ-ω gonio scan, and the other is pole figure configuration, *i.e.* rotating on the axis normal to the substrate surface (azimuthal angle ϕ) and rotating on the intersecting axis between the substrate surface and the X-ray path plane (azimuthal angle ψ), and fixing 2θ and ω at the specific crystal plane, as shown in Figure 2. In pole figure configuration, the orientation of the specific crystal planes can be determined.

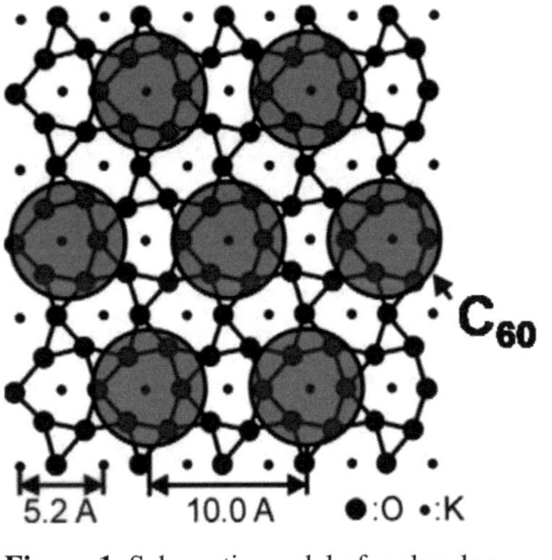

Figure 1. Schematic model of molecular arrangement of C_{60} on mica (001) surface.

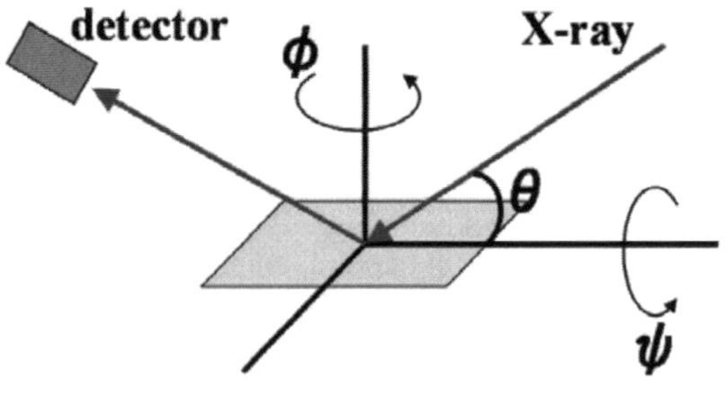

Figure 2. Pole figure configuration in XRD measurement.

3. Results and Discussion

Figure 3 shows the XRD patterns of undoped C_{60} films on glass and mica (001) substrates. For undoped C_{60} films on glass, the observed diffraction peaks correspond with polycrystalline C_{60} phase with fcc lattice, though these diffraction peaks are rather broad. In contrast, undoped C_{60} films on mica (001) shows only sharp (111)-related diffraction peaks, suggesting good crystal quality.

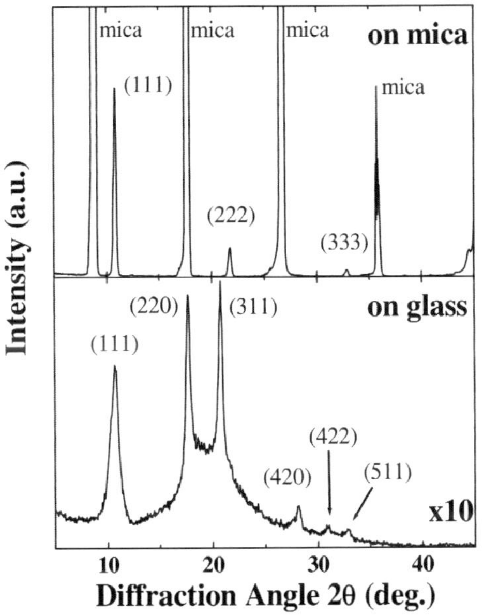

Figure 3. XRD patterns of undoped C_{60} films on glass and mica (001) substrates.

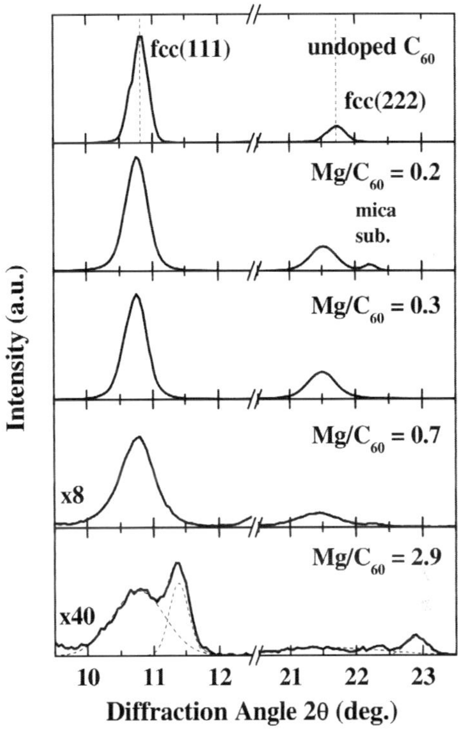

Figure 4. XRD patterns of Mg-doped C_{60} (Mg/C_{60} molar ratio = 0~2.9) grown on mica (001).

Figure 4 shows the XRD patterns of Mg-doped C_{60} (Mg/C_{60} molar ratio = 0~2.9) grown on mica (001). In Mg-doped C_{60} with Mg/C_{60} molar ratio of 0~0.7, fcc (111)-related sharp diffraction peaks are observed and the lattice parameter is expanded from a=14.14Å for undoped C_{60} to 14.29Å by Mg incorporation, though these peaks become a little broader with increasing Mg concentration. While in Mg-doped C_{60} with Mg/C_{60} molar ratio of 2.9, these peaks considerably weakened, suggesting less crystal quality, and fcc (111) diffraction peak splits into two peaks. This result suggests the possibility of phase separation or anisotropic lattice parameter change. Recently, formation of rhombohedral 2-dimensional polymer of Mg_xC_{60} (x=4,5) is reported [4, 5]. If we attribute these peaks to rhombohedral, the lattice parameters are a=b=9.44Å, c=24.61Å. From the calculated a value, it is indicated that the distance between adjacent C_{60} molecules of Mg-doped layer is shorter than that of pure C_{60} film. It is suggested that C_{60} polymer sheets are possibly formed in high Mg concentration region (Mg/C_{60} molar ratio > 2).

In the low Mg concentration region (Mg/C_{60} molar ratio < 1), the relation of crystal orientation between the mica substrate and Mg-doped C_{60} film was investigated by the XRD with pole figure configuration. Figure 5 shows the pole figure mapping of C_{60} {111} planes and mica {$11\bar{2}$} planes of Mg-doped C_{60} film with Mg/C_{60}=0.8. The center of the circle corresponds to ϕ=0 and ψ=0, and means the normal direction of the sample surface. From this figure, it is clearly seen that C_{60} {111} planes have same symmetry with mica {$11\bar{2}$} planes. Thus, we can conclude that Mg-doped C_{60} film is grown epitaxially on mica (001) substrate in the low Mg concentration region (Mg/C_{60} molar ratio < 1).

Figure 6 shows Mg concentration dependence of the conductivity in comparison between glass and mica substrates. For the mica substrate, the conductivity was measured in the low Mg concentration region where the epitaxial growth was obtained. The conductivity increases with increasing Mg concentration in both substrates. Furthermore, significant increase in conductivity can be seen in Mg-doped C_{60} film on mica substrate. It is considered that such significant increase is caused by the drastic improvement of crystal quality realized by epitaxial growth, as shown in Figure 3. This result may indicate the significant increase of carrier mobility. The difference in conductivity between mica and glass substrates decreases with increasing Mg concentration. This tendency may be caused by the crystal quality degradation with Mg-doping in the films on mica substrate, as seen in fig. 4.

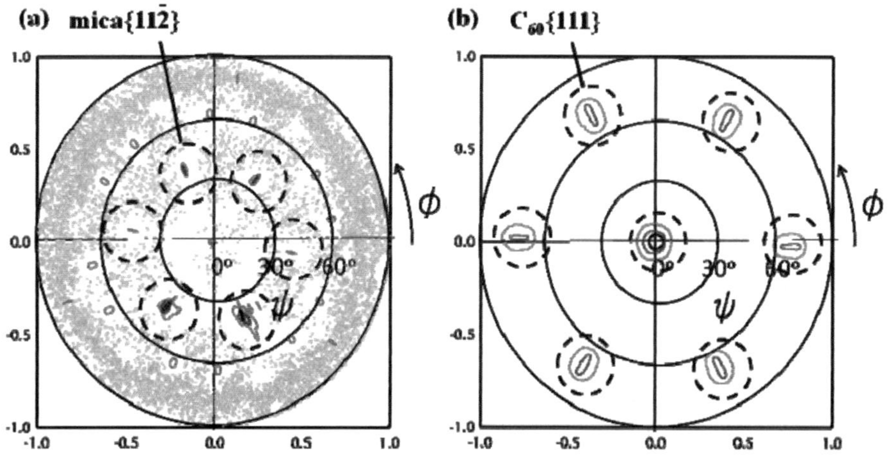

Figure 5. Pole figure mapping of (a) mica {11$\bar{2}$} planes and (b) C_{60} {111} planes of Mg-doped C_{60} film with Mg/C_{60}=0.8.

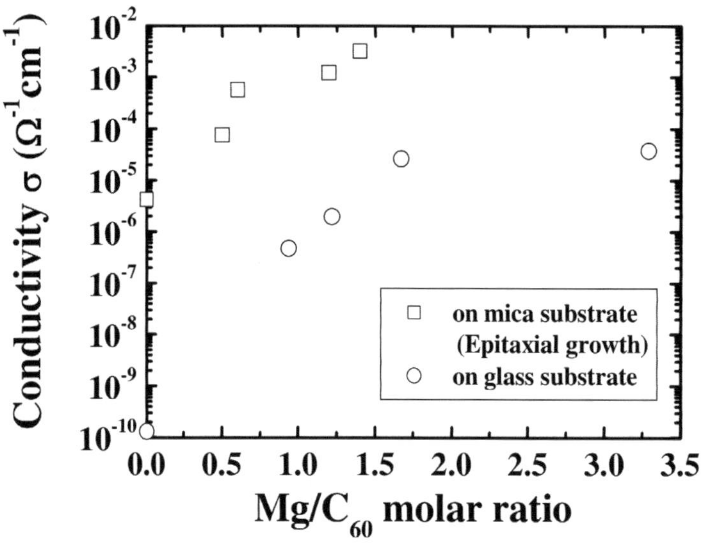

Figure 6. Mg concentration dependence of the conductivity in comparison between glass and mica substrates.

4. Conclusion

The epitaxial growth of Mg-doped C_{60} films was investigated. It was found that the epitaxial growth of Mg-doped C_{60} film was enabled by using mica (001) substrate in the low Mg concentration region (Mg/C_{60} molar ratio < 1). The crystal quality of the epitaxial Mg-doped C_{60} film was improved drastically in compared with micro-crystalline film on glass substrate. Such drastic improvement of crystal quality in the epitaxial films resulted significant increase in conductivity. This result may indicate the significant increase of carrier mobility. In addition, the crystal structure change was observed in high Mg concentration region (Mg/C_{60} molar ratio > 2). The C_{60} polymer sheet formation was suggested.

Acknowledgements
This work was supported in part by the MEXT as a Private University High-Tech. Research Center Program 2001-2008.

References

[1] Gupta R P and Gupta M 1994 *Physica C* **219** 21
[2] Chikamatsu M, Taima T, Yoshida Y, Saito K and Yase K 2004 *Appl. Phys. Lett.* **84** 127
[3] Chikamatsu M, Nagamatsu S, Taima T, Yoshida Y, Sakai N, Yokokawa H, Saito K and Yase K 2004 *Appl. Phys. Lett.* **85** 2396
[4] Borondics F, Oszlányi G, Faigel G and Pekker S 2003 *Solid State Commun.* **127** 311
[5] Quintavalle D, Borondics F, Klupp G, Baserga A, Simon F, Jánossy A, Kamarás K and Pekker S 2008 *Physical Review B* **77** 155431

Photopolymerization of C_{60} Crystal Synthesized from Organic Solution

Y Iio, K Kurihara, F Matsuyama, R Nokariya, N Iwata and H Yamamoto
Departments of Electronics and Computer Science, College of Science and Technology
Nihon University, Funabashi, Japan

E-mail: hyama@ecs.cst.nihon-u.ac.jp

Abstract. We have been developing a novel process to synthesize C_{60} polymers as superior functional materials. Polymerization reactions of C_{60} precipitated from organic solution were investigated under Free Electron Laser (FEL) irradiation. The two processes: a liquid-liquid interfacial precipitation (LLIP) method and an evaporation technique from supersaturated solution, were studied with a respect of crystal structure and/or Raman spectrum for the precipitates before and after FEL irradiation. It was found that the photopolymerization reaction took place effectively in the LLIP precipitate because of its highly packed molecular crystal structure.

1. Introduction

Since C_{60} polymers were found first by phototransformation [1], many studies have been done for many C_{60} polymerization processes. It can be expected that C_{60} polymers become marvelous super-materials accompanying with very low-density, flexibility, and ultra-hardness etc. We have studied photo-assisted polymerization processes for synthesis of 3-dimensional polymers from pressed C_{60} powder by irradiation of Free Electron Laser (FEL) which has the features of wavelength variability and extremely sharp pulse width [2-5]. The specimen after FEL irradiation revealed a characteristic red-shift of Raman $A_g(2)$ peak. The obtained results suggested that FEL irradiation with wavelength of 400~500nm was effective for photo-polymerization reactions and/or smaller C_{60} intermolecular distance was more proper to promote polymerization reactions.

The purposes of this work are to study the crystal structure of C_{60} precipitates grown by a liquid-liquid interfacial precipitation (LLIP) method [6] and an evaporation technique from supersaturated solution using organic solvents: toluene and m-xylene, and to investigate FEL irradiation effects for the obtained precipitates with a respect of crystal structure and/or Raman spectrum. The result of the precipitates is discussed compared with that of photopolymers synthesized from C_{60} pressed powder.

2. Experimental

Fig. 1 shows the schematic process of the evaporation technique from supersaturated solution. Commercial C_{60} powder (99.5%) was dissolved to organic solution with 50ml test tube (inner diameter 35×height 115mm, surface area 491mm^2). At first C_{60} was dissolved into toluene or m-xylene. After an ultrasonic dispersion was done for 5 minutes to prepare saturated solution, C_{60} (0.1g) was putted into the saturated solution (10ml) to make supersaturated solution. Then SiO_2 substrates attached to fluorine strings were soaked into the supersaturated solution and were preserved at 12°C in a refrigerator to suppress evaporation speed. The substrates appeared out of the solution after about 7 days.

The LLIP process conditions were as following. Isopropyl alcohol (20ml) was added cautiously into the C_{60} supersaturated solution (10ml). Then it was also left at 12 °C for 7 days in the refrigerator.

The used FEL has the following features: a variable wavelength from 800nm to 3000nm as a fundamental and a macro-pulse (@20μs) containing very short micro-pulses (several hundreds fs) [7] . As reported 400~500 nm-FEL irradiation effectively promoted C_{60} polymerization reactions. The 500nm FEL, which was the 3rd harmonics of fundamental 1500nm FEL, was irradiated through a band pass filter for 60 min and pressure was 10^{-4} Pa. The power of the fundamental FEL was *ca.* 3.0 mJ/pulse cm^2.

The crystal structure of the obtained specimens was analyzed by X-ray diffraction (XRD) (Rigaku International Corp.: RAD-C). The Micro-Raman spectroscope (Kaiser Corp.: Holo Lab 5000R) was used to investigate molecular vibrations. The Raman $A_g(2)$ mode of C_{60} molecules in the vicinity of 1469cm^{-1} is defined as contracting vibration of pentagon rings. It is known that the $A_g(2)$ pentagonal pinch mode peak is broaden and shifts to the lower-energy side; e.g. the photo-polymerization results in a peak red-shift by approximately 10cm^{-1}. Then the $A_g(2)$-derived mode has been therefore noted as an index of the polymerization reaction progress.[8]

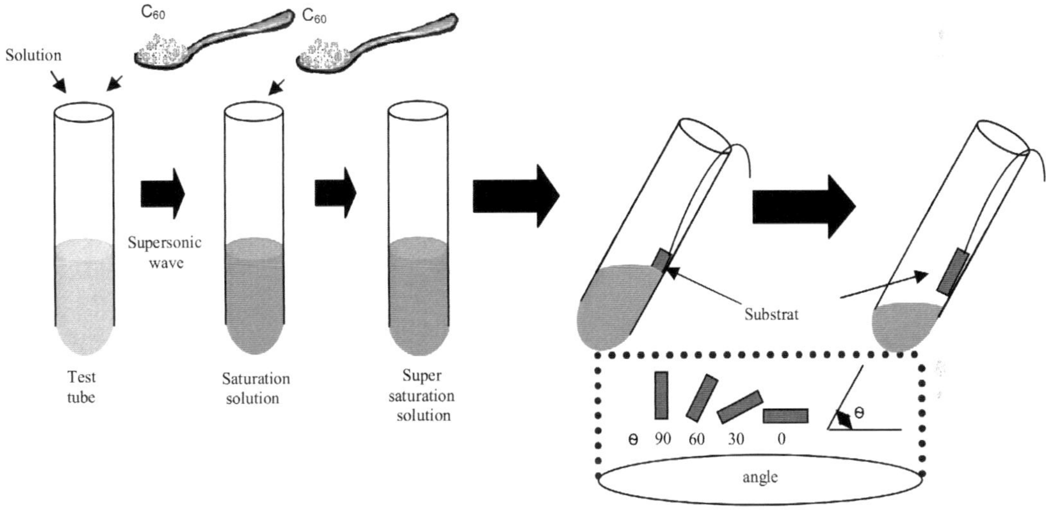

Fig. 1. The schematic process of the evaporation technique from supersaturated solution.

3. Results and discussion

Fig. 2 shows the optical microscope images of the evaporation method specimens grown from supersaturated toluene solution. Precipitates with the rectangular shape grew on the substrate which was set up at the angle of 60°. The shape of the precipitate was plate-like rather than needle-like probably because blemishes or scratches became nuclei of growth. The precipitates hardly grew on the substrates set up at the angles of 0° and 30° and short needle-like precipitates on the substrate at the angle of 90°. The contact area between the substrate and solution was an important factor for the growth of precipitates during evaporation. This is the reason why the shape or the amount of the precipitates drastically changed according to the substrate set up angle. Fig. 3 shows the optical microscope images of the evaporation method specimens grown from supersaturated m-xylene solution. Characteristic needle-like precipitates were obtained, not depending on the substrate set up angle. Especially very long precipitates grew up at the angle of 60°. It should be noticed that the crystal growth using the m-xylene solution took place preferably with a hexagonal close-packed (hcp) phase and the obtained shape of the precipitate became needle-like as the result [9].

Fig. 2. Optical microscope images of the evaporation method specimens grown from supersaturated toluene solution on the substrate set up angle of 0° (a), 30° (b), 60° (c) and 90° (d).

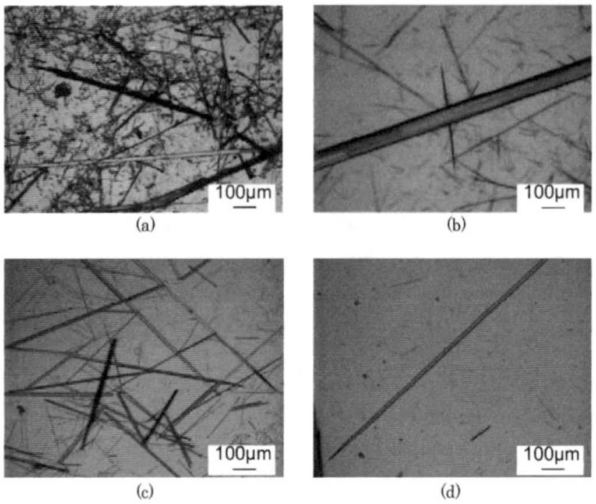

Fig. 3. Optical microscope images of the evaporation method specimens grown from supersaturated m-xylene solution on the substrate set up angle of 0° (a), 30° (b), 60° (c) and 90°(d).

Fig. 4 shows the optical microscope images of the LLIP method specimens obtained by using toluene and m-xylene. The needle-like precipitates with comparatively long size grew on all over surfaces of the substrate in toluene solution. In the case of m-xylene solution obtained precipitates were also needle-like, but their length was short, at most about 100μm. Since the amount of C_{60} dissolved for toluene was larger than that for m-xylene, comparatively long precipitates of the maxium of about 10mm and/or much precipitates were grown up from m-xylene solution.

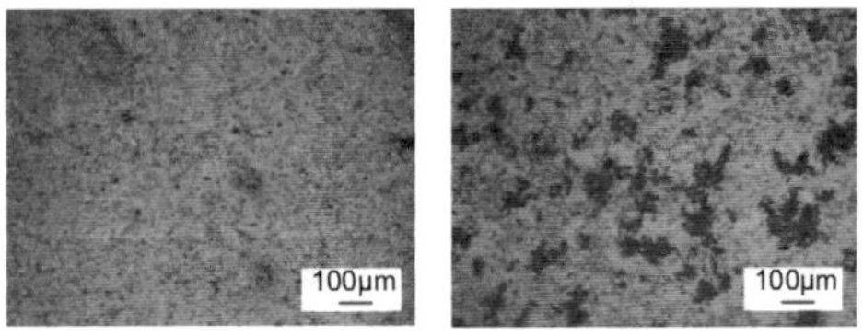

(a) (b)

Fig. 4. Optical microscope images of the LLIP method specimens obtained by using toluene (a) and m-xylene (b).

Table1. The fcc lattice d-space value of the typical samples. The upper and the lower values are the result before and after FEL irradiation, respectively.

fcc plane	Evaporation of supersaturated solution		LLIP method		C_{60} powder
	Toluene	m-xylene	toluene	m-xylene	-
	d-space of the lattice plane [Å]				
(111)	8.3 8.2	8.3 -	8.3 8.1	8.2 8.1	8.3 8.2
(220)	5.1 5.1	5.1 5.1	5.1 5.0	5.0 5.0	5.1 5.1
(311)	4.4 4.3	4.4 4.4	4.4 4.3	4.3 4.3	4.4 4.4

The structures of the specimens obtained were analyzed from XRD patterns. The d-space values corresponding to the face centered-cubic (fcc) phase are summarized in Table 1. The typical XRD pattern of the toluene and m-xylene LLIP specimen before and after FEL irradiation is shown in Figs. 5, 6 and Figs. 7, 8, respectively. As the general tendency the intermolecular distances of the precipitates from organic solution were smaller than those of C_{60} powder pressed under 600 MPa. In the both evaporation method specimens from the supersaturated toluene and m-xylene, the intermolecular distance changed not so much by FEL irradiation. On the other hand, the intermolecular distance of the LLIP specimens before irradiation was smaller than that of C_{60} powder. And the lattice constants of the fcc phase obviously decreased by FEL irradiation. The higher dense

C_{60} packing was attained by using the LLIP method probably because of large internal compressive stress in the precipitating process via liquid-liquid interfacial region. Especially the m-xylene LLIP specimen revealed a large hallo pattern, which indicated the existence of large crystalline distortion. M. Sathish et al.[9] reported that the LLIP precipitate showed the tendency to form the hcp phase in the solution though its phase was not so stable in air. The result of m-xylene specimen was the case. The similar mechanism took place in the specimens and additional diffraction peaks assigned to the hcp phase appeared in Figs. 6 and 8. The reason why the intermolecular distortion of the precipitate from toluene was comparatively smaller than that from m-xylene was not clear, but which may be related to the number of hydroxyl terminals.

After FEL irradiation the hallo diffraction pattern appeared also in the both LLIP specimens and the lattice space became almost same. The polymerized bond length between C_{60} molecules is shorter by about 10% than the average intermolecular length of a monomer state. Since the average lattice constants expand in amorphous structure as a general argument, the polymeric C_{60} bonds result in the expanding distortion of the crystal structure as proceeding polymerization. The similar discussion appeared in elesewhere[8]. So the conflicting two mechanisms worked in the polymerization process. It was understood that the large decrease in the lattice space observed may be caused by polymerized bonding effects were so large.

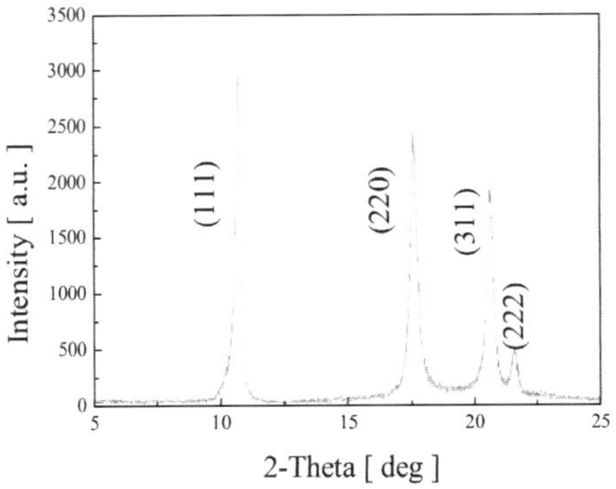

Fig. 5. XRD pattern of the toluene LLIP specimen before FEL irradiation. The index appeared in the figure means the each planes from the fcc phase.

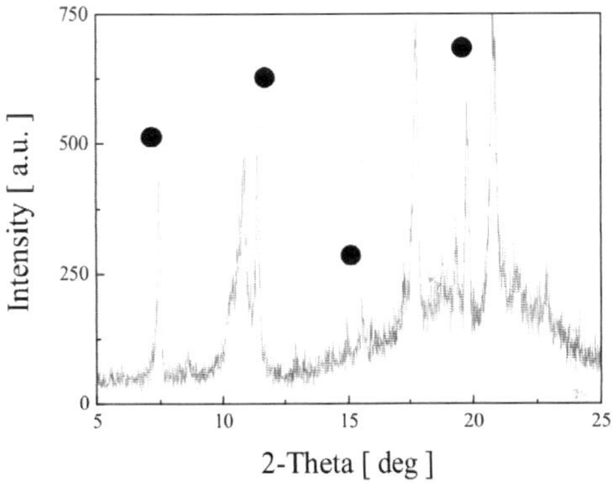

Fig. 6. XRD pattern of the m-xylene LLIP specimen before FEL irradiation. The index appeared in the figure means the each planes from the fcc phase and the dots correspond to the diffraction peaks assigned as the hcp phase.

Journal of Physics: Conference Series **159** (2009) 012019 doi:10.1088/1742-6596/159/1/012019

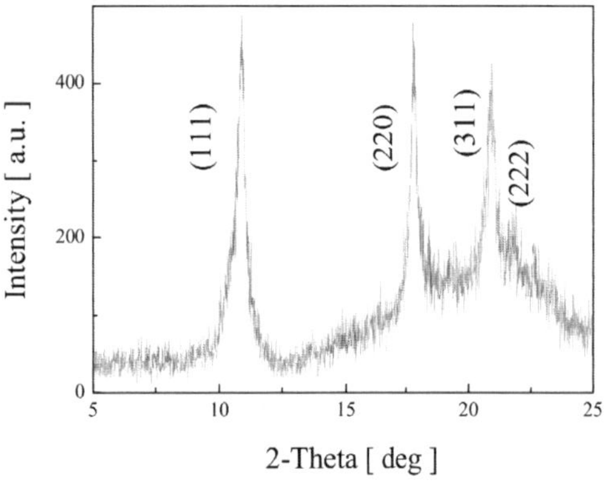

Fig. 7. XRD pattern of the toluene LLIP specimen after FEL irradiation. The index appeared in the figure means the each planes from the fcc phase.

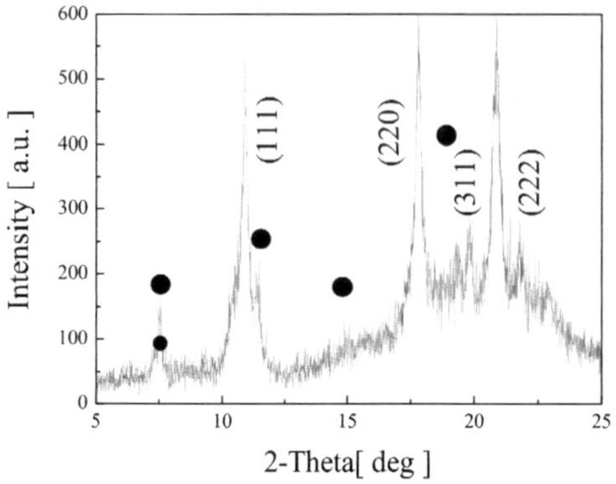

Fig. 8. XRD pattern of the m-xylene LLIP specimen after FEL irradiation. The index appeared in the figure means the each planes from the fcc phase and the dots correspond to the diffraction peaks assigned as the hcp phase.

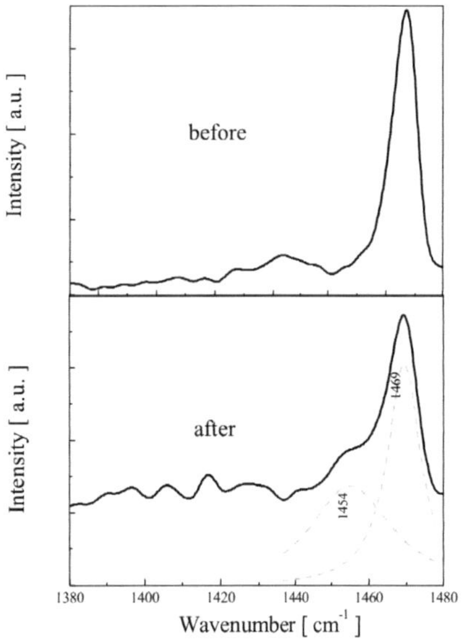

Fig. 9. Raman spectrum of the toluene LLIP specimen before and after FEL irradiation. The dotted lines mean the decomposed peaks.

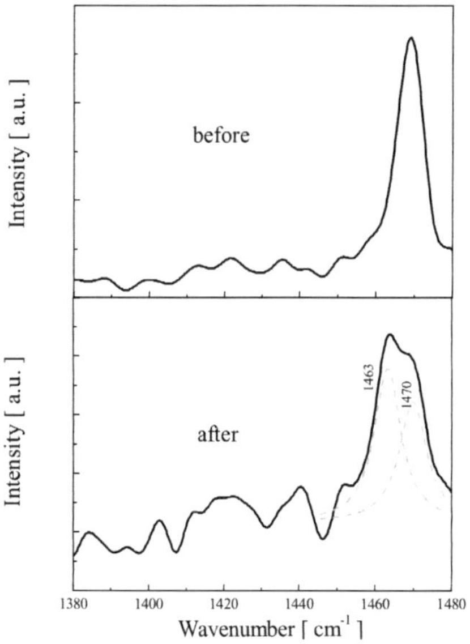

Fig. 10. Raman spectrum of the m-xylene LLIP specimen before and after FEL irradiation. The dotted lines mean the decomposed peaks.

The Raman spectra of the pressed powder and/or the toluene and m-xylene LLIP specimens are shown in Figs. 9-10, comparing with the each result before and after FEL irradiation. The peak at 1469cm^{-1} identified as $A_g(2)$-derived mode of C_{60} was confirmed in all specimens. In the case of the pressed powder obvious red-shift of $A_g(2)$ peak was not observed even after FEL irradiation because not so high pressure, 600 MPa resulted in small changes of intermolecular distances. The main peak at 1469cm^{-1} and the small red-shifted 1454 cm^{-1} peak, which indicated partial strong photo-polymerization, were observed in the toluene LLIP specimen after FEL irradiation. On the other hand, the main peak shifted to *ca.* 1463 cm^{-1} in the m-xylene LLIP specimen. Furthermore it revealed the characteristic peak tails in the lower energy side, which suggested the promoted photo-polymerization. So the polymerization reactions took place more effectively in the m-xylene specimen rather than in the toluene specimen. It was thought that the difference of the polymerization was caused by the different crystalline structure of the each obtained precipitate as discussed in the results of XRD. These results strongly suggested that the photo-polymerization was promoted as C_{60} intermolecular distances decreased.

4. Conclusion

The C_{60} precipitates were grown from organic solution by two methods: the evaporation from supersaturated solution and the LLIP method. The polymerization reaction by FEL irradiation was investigated by the crystal structure and/or Raman spectroscopic analysis. It was found that the intermolecular distances decreased and the polymerization was promoted especially in the LLIP specimens. Conclusively the LLIP method was useful to get raw materials for the effective synthesis of C_{60} polymers.

Acknowledgments

The authors are grateful for the financial support of the Grant-in-Aid for the Futaba Foundation. The authors are also grateful for the financial support of an Interdisciplinary General Joint Research Grant (Nihon University). This work was supported by the Ministry of Education, Culture, Sports, Science and Technology through a Scientific Grant-In-Aid (No. 20360145).

References

[1] Rao A M, Zhou P, Wang K A, Hager G T, Wang Y, Lee W T, Bi X X, Eklund P C, Cornett D S, Duncan M A and Amster I J 1993 *Science* **259** 955
[2] Ando A, Nokariya R, Koyaizu R, Iwata N and Yamamoto H 2007 *Trans. Mate. Res. Jpn.* **32** 1251
[3] Iwata N, Ando S, Nokariya R and Yamamoto H 2007 *Jpn. J. Appl. Phys.* **47** 1412
[4] Yamamoto H, Iwata N, Hashimoto R and Ando S 2007 *Appl. Surf. Sci.* **253** 7977-80
[5] Iio Y, Ando S, Nokariya R, Iwata N and Yamamoto H 2008 *The 34th Commemorative Fullerene- Nanotubes General Symposium* 3P-49 199
[6] Minato J and Miyazawa K 2005 *Sci. Tech. Adv. Mate.* **6** 388
[7] Hayakawa Y, Sato I, Hayakawa K, Tanaka T, Yokoyama K, Kanno K, Sakai T, Ishikawa K, Nakao K and Hashimoto E 2003 *Nucl. Instrum. Methods Phys. Res. Sect.* A **507** 404
[8] Eklund P C and Rao A M 1999 *Fullerene Polymers and Fullerene Polymer Composites* (Springer)
[9] Sathish M, Miyazawa K and Wakahara T 2008 Preparation of Fullerene (C_{60}) Nanosheets by Liquid-liquid Interfacial Precipitation Method *The 35th Commemorative Fullerene-Nanotubes General Symposium* 1-15 199

Crystallinity and properties of C_{60} nanotubes improved by annealing and alcohol-soaking

K Naito and K Matsuishi

Institute of Materials Science, University of Tsukuba, Tsukuba, Ibaraki 305-8573, Japan

E-mail: kiyoto@bk.tsukuba.ac.jp

Abstract. Well-uniformed C_{60} nanotubes were grown at -20 °C with irradiation of red light using C_{60}-saturated pyridine solution and isopropyl alcohol by a liquid-liquid interfacial precipitation method without ultrasonic pulverization. We attempted to improve their crystallinity by two post-treatments; thermal annealing and alcohol-soaking. The crystallinity of as-grown and dried C_{60} nanotubes, which was poor due to the evaporation of solvent molecules from crystals in the drying process, was improved by annealing around 220 °C for 5 hours in vacuum. Dramatic improvement of crystallinity of as-grown samples was achieved by soaking into methanol and then drying in air. Raman, infrared and X-ray diffraction results suggest that the methanol-soaked samples exhibit a solvated tetragonal structure. The crystallinity improved by methanol-soaking did not degrade after removal of methanol molecules from samples by thermal annealing. Photo-polymerization of the structurally-improved C_{60} nanotubes was examined to investigate an effect of crystallinity on the polymerization kinetics.

1. Introduction

Fullerene crystals composed of C_{60} molecules have been synthesized [1-3]. Miyazawa et al. have synthesized single crystalline whiskers with diameters smaller than 1 μm and a length of more than 100 μm by a liquid-liquid interfacial precipitation method [4, 5]. A C_{60} nanowhisker with a diameter of 470 nm is flexibly bent with a curvature radius of 38 μm [5]. Recently, fullerene nanowhiskers with hollow structures, which are called as fullerene nanotubes, have been fabricated [6, 7]. They have wall structures made of fullerene molecules. Both fullerene nanowhiskers and nanotubes have drawn attention as a quasi-one dimensional nanomaterial due to their potential application to nanotechnology.

Fullerene nanowhiskers and nanotubes crystallize to form solid solvates with solvents. Especially, solvated structures of C_{60} nanotubes are not air-stable and rapidly loose solvent molecules from the crystals by evaporation. This follows a rapid change in the structure, and consequently the crystallinity of the C_{60} nanotubes is degraded. Such a structural degradation affects the properties of C_{60} nanotubes. Hence, the preparation of C_{60} nanotubes with good quality has been desired for evaluation of their intrinsic properties.

In this study, we synthesized C_{60} nanotubes by a liquid-liquid interfacial precipitation method, and attempted to improve their crystallinity by thermal annealing in vacuum and alcohol-soaking processes. We characterized samples by X-ray diffraction (XRD), Raman scattering and Infrared (FTIR) measurements, and discussed their structures. Photo-polymerization of structurally-improved C_{60} nanotubes was then examined to investigate an effect of crystallinity on the polymerization kinetics.

2. Experimental

C_{60} nanotubes were prepared by a liquid-liquid interfacial precipitation method using C_{60}-saturated pyridine solution and isopropyl alcohol (IPA). Our typical procedure was as follows. 3 mL of pyridine solution saturated with C_{60} was put into a glass bottle and then 27 mL of IPA was gently added. The bottle was capped and kept at -20 °C for more than one week. To obtain suitable diffusion at the interface, the glass bottle was shaken by hand. In our procedure, the ultrasonic pulverization which was reported necessary after the formation of liquid-liquid interface [6] was not executed. Since the photo-assisted growth of C_{60} nanowhiskers was reported [8], we adapted this technique to the growth of C_{60} nanotubes as well using red light from a Light Emitting Diode (LED) with the center wavelength of 627 nm. We attempted to improve the crystallinity of as-grown samples by thermal annealing and alcohol-soaking. The thermal annealing was performed in vacuum (about 1×10^{-3} Pa by a diffusion vacuum pump system) at temperatures between 180 and 400 °C. Our typical alcohol-soaking procedure was as follows. As-grown samples were taken out of a glass bottle and soaked immediately into alcohols using a pipette for several hours. Then, the samples were dried completely in air. We examined several alcohols with different molecular weights; methanol, ethanol and 1-propanol for our soaking experiment.

Morphological observations were performed using an ordinary optical microscope, SEM (JEOL, JSM-5510) and TEM (JEOL, JEM-2010). To evaluate residual solvent molecules in samples, infrared absorption measurements were performed for pristine C_{60} nanotubes and C_{60} nanotubes after soaking into various alcohols using a FTIR apparatus (VM7, JASCO). XRD profiles were obtained using an X-ray diffractometer with CuK_{α} radiation. Raman scattering spectra were measured in vacuum (about 1×10^{-4} Pa by a turbo molecular pump system) using JASCO TRS-600 and an Ar^{+} laser of the wavelength of 514.5 nm to investigate the influence of solvent molecules to the molecular vibrations of C_{60} as well as to elucidate the photo-induced polymerization kinetics of C_{60} nanotubes.

3. Results and discussion

3.1. Sample preparation

When C_{60} nanotubes were sufficiently grown, bundles of nanotubes with several millimeters in length were easily recognized by eye as shown in figure 1(a). An optical micrograph of the sample in air is shown in figure 1(b). Figures 2(a) and 2(b) show a TEM image and a SEM image, respectively, of the C_{60} nanotubes. The outer diameter is about 600 nm in figure 2(a) and about 3 μm in figure 2(b). Our observations revealed that the bundles consisted of nanotubes of about several hundred nanometers in outer diameter on average. We found that C_{60} nanotubes could be synthesized without ultrasonic pulverization, although it had been thought necessary after the formation of liquid-liquid interface.

By the photo-assisted growth using the LED light (wavelength; 627nm), C_{60} nanotubes grew up to more than 20 mm in length. The observation suggests that the growth rate of C_{60} nanotubes can be promoted by the red light illumination, as observed previously for C_{60} nanowhiskers [8]. There was a size distribution in the outer diameter from about 250 nm to about 1500 nm. The average outer diameter was about 700 nm. The size distribution tends to become narrower with a lower growth temperature.

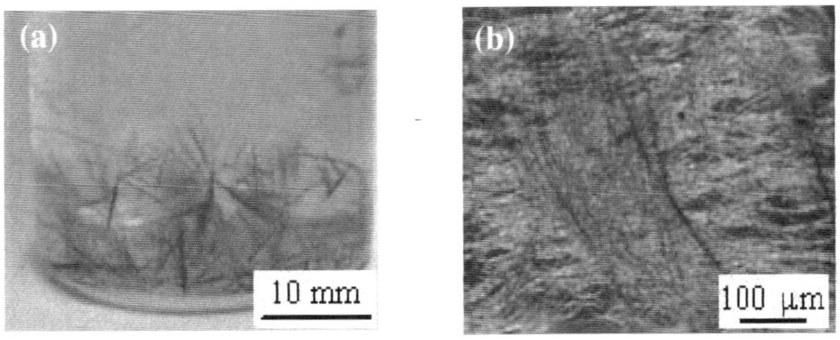

Figure 1. (a) Photograph of C_{60} nanotubes grown in a glass bottle by forming liquid-liquid interface between C_{60}-saturated pyridine and IPA, and (b) optical micrograph of the C_{60} nanotubes taken into air from the bottle.

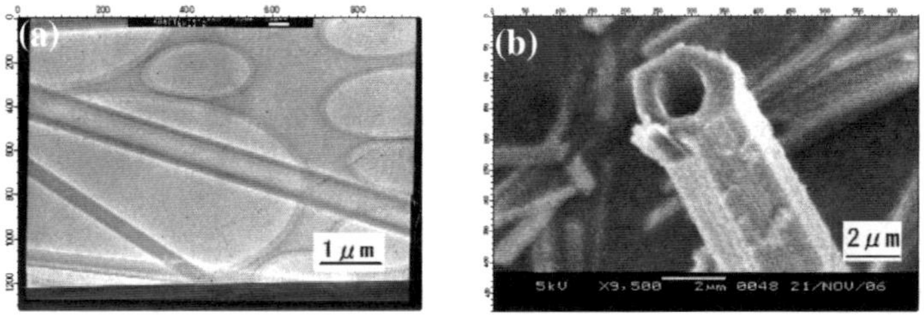

Figure 2. (a) TEM image, and (b) SEM image of the C_{60} nanotubes grown at -20 °C under illumination of red light from a LED (wavelength; 627 nm).

3.2. Improvement of crystallinity by annealing in vacuum and alcohol-soaking

Crystalline fullerene solids precipitated from solutions sometimes show morphologies different from the ordinary face-centered cubic (fcc) packing of fullerene molecules [9]. Fullerene molecules should be solvated when they are in solution. For fullerene nanotubes, they are known to crystallize in the solvated structure of hexagonal symmetry with cell dimensions a = 1.541 nm and c = 1.00 nm and to change into a fcc structure by the evaporation of solvent molecules after drying in air [10]. As a result, the crystallinity of as-grown and dried C_{60} nanotubes is poor. Our as-grown samples exhibited a solvated hexagonal structure with a = 1.540 nm and c = 0.99 nm right after taking into air from the

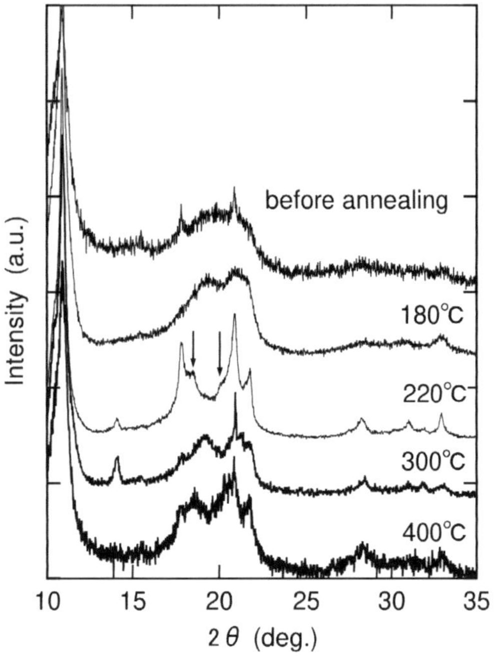

Figure 3. Annealing temperature dependence of XRD profiles of C_{60} nanotubes. Annealing was carried out in vacuum for 5 hours at the temperature indicated. The uppermost profile is for an as-grown sample just after drying in air.

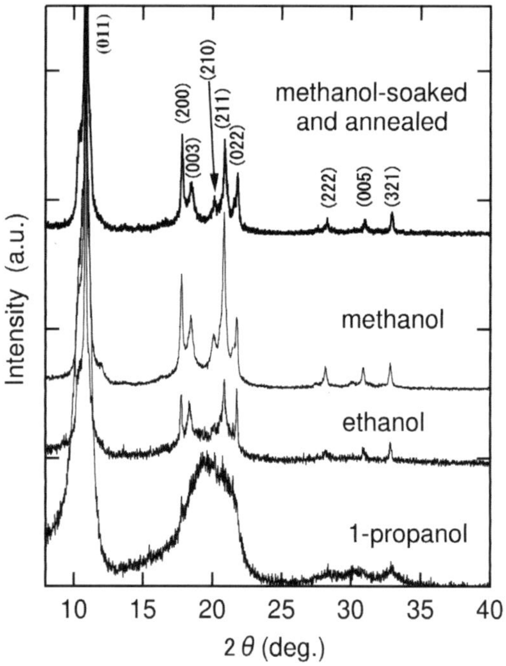

Figure 4. XRD profiles of C_{60} nanotubes soaked into alcohols and then dried in air. The uppermost profile is for the sample soaked into methanol and then annealed at 260 °C for 5 hours.

solution, and changed in 10 min in air to a poor crystalline structure as realized in the uppermost XRD profile in figure 3. FTIR spectra after dried in air showed very small peaks of intramolecular vibrational modes of pyridine (at 704 cm^{-1}) and IPA (at 954 and 1110 cm^{-1}) besides four intense peaks of C_{60} molecules (at 527, 576, 1182 and 1428 cm^{-1}). This suggests the presence of a small amount of pyridine and IPA molecules in the as-grown/dried samples.

To improve the crystallinity of C_{60} nanotubes, we performed thermal annealing in vacuum at temperatures between 180 and 400 °C at intervals of 20 °C. Figure 3 shows XRD profiles for the C_{60} nanotubes annealed in vacuum at 180, 220, 300 and 400 °C together with that of the as-grown /dried sample. It is found that annealing around 220 °C in vacuum made the diffraction peaks the sharpest, resulting from structural relaxation and removal of residual solvent molecules from samples. The samples annealed in vacuum seem to exhibit the same structure as pristine C_{60} powder (*i.e.*, an fcc structure). However, the peaks that cannot be indexed by the fcc structure were observed as indicated by arrows in figure 3. The observation suggests the lowering of the symmetry of the fcc structure. We indexed the diffraction peaks by a tetragonal structure and obtained the lattice constants, a = 0.995 nm and c = 1.437 nm.

Figure 4 shows the XRD profiles for the C_{60} nanotubes after alcohol-soaking followed by drying in air. Using methanol of the smallest molecular weight, the diffraction peaks became the sharpest among these alcohols. They can be indexed by a tetragonal structure, a = 1.001 nm and c = 1.446 nm. It should be noticed that by the methanol-soaking the diffraction peaks became sharper than those of the samples annealed at 220 °C shown in figure 3. Figure 5 shows the FTIR spectra of the C_{60} nanotubes with the alcohol-soaking treatments. For the methanol-soaked sample, the absorption peak which originated from a vibrational mode of methanol was observed as indicated by an arrow in figure 5. Besides, very weak vibrational modes of pyridine and IPA are seen after the alcohol-soaking. This suggests that the solvents are removed from the solvated hexagonal structure of as-grown sample and methanol molecules diffuse into the C_{60} lattice to form a solvated (C_{60}-methanol) tetragonal structure. We examined the soaking time dependence of Raman spectra for methanol-soaked samples. Figure 6 shows a change in the A_g (2) pentagonal pinch mode of C_{60} with increasing the soaking time. The A_g (2) pentagonal pinch mode shifted to lower frequencies from 1467.9 to 1465.7 cm^{-1} due to interaction between C_{60} and methanol molecules, as the soaking time increased from 0.5 to 4 hours. Simultaneously, the XRD peaks became shaper as the soaking proceeded. These XRD, FTIR and

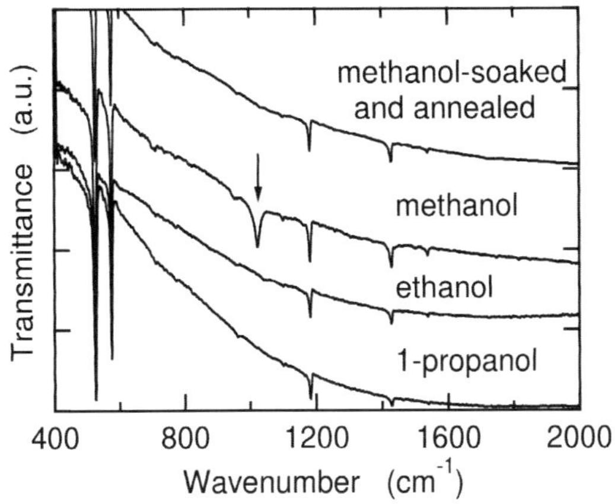

Figure 5. FTIR spectra of the C_{60} nanotubes soaked into alcohols and then dried in air. The uppermost spectrum is for the sample soaked into methanol and then annealed at 260 °C for 5 hours.

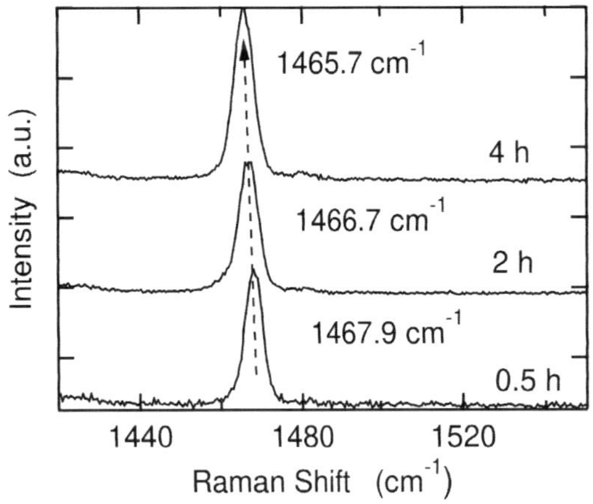

Figure 6. Soaking time dependence of Raman spectra of the C_{60} nanotubes with methanol-soaking treatments. Experiment was carried out at 77 K in vacuum to prevent samples from heating which may induce graphitization.

Raman results indicate that since the C_{60} nanotubes are solvated with methanol to form a stable tetragonal structure, methanol molecules remain in the C_{60} lattice stably in air at room temperature after methanol-soaking.

Our significant finding is that the improved crystallinity by methanol-soaking did not degrade after removal of methanol molecules from samples by annealing as seen in figure 4. The absorption peak from methanol disappeared after annealing at 260 °C for 5h, as shown in the uppermost spectrum in figure 5, indicating that the methanol molecules were certainly removed from the C_{60} nanotubes. We found that the lattice constants were slightly changed to a = 0.998 nm and c = 1.442 nm after the annealing on the methanol-soaked sample.

3.3. Photo-induced transformation

We have measured the Raman scattering spectra of C_{60} nanotubes under laser irradiation of the 514.5 nm line from an Ar^+ laser to investigate an effect of crystallinity on the photo-polymerization of C_{60} nanotubes.

Figure 7 shows the relative intensity of the $A_g(2)$ pentagonal pinch mode to the polymer mode which appears around 1461 cm^{-1} due to polymerization of C_{60} molecules as a function of laser irradiation time for methanol-soaked and non-soaked samples. Both samples were annealed at 260 °C for 5 hours in vacuum before laser irradiation. The time constants for the photo-polymerization of the C_{60} nanotubes with and without soaking are 6.6 and 10 min, respectively. This indicates that the C_{60} nanotubes soaked into methanol are easier to polymerize than non-soaked samples. We infer that the improvement of crystallinity in C_{60} nanotubes promotes photo-polymerization, since the structurally-improved C_{60} nanotubes by methanol-soaking followed by annealing has better crystallinity than the annealed C_{60} nanotubes without alcohol-soaking as mentioned above from the XRD results.

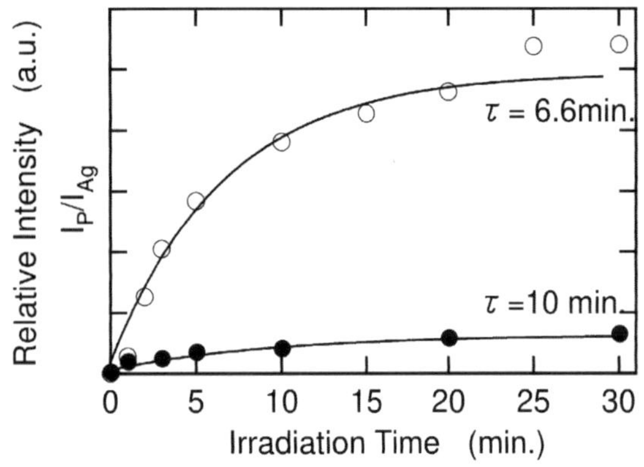

Figure 7. Changes in the relative Raman intensity of polymer mode to $A_g(2)$ pentagonal pinch mode with laser irradiation time. Laser irradiation and Raman experiments were carried out at room temperature in vacuum. Open circles denote data for structurally-improved C_{60} nanotubes by methanol-soaking followed by annealing, and solid circles for annealed C_{60} nanotubes without alcohol-soaking.

4. Conclusion

We synthesized C_{60} nanotubes using C_{60}-saturated pyridine solution and isopropyl alcohol by a liquid-liquid interfacial precipitation method. Well-uniformed C_{60} nanotubes were grown successfully at -20 °C without ultrasonic pulverization. Using a photo-assisted growth technique with LED light of wavelength 627 nm, C_{60} nanotubes grew up to more than 20 mm in length. We succeeded to improve their crystallinity by two post-treatments; thermal annealing and alcohol-soaking. The C_{60} nanotubes annealed at 220 °C for 5 hours in vacuum made diffraction peaks the sharpest among the samples annealed at temperatures between 180 and 400 °C. This improvement is due to structural relaxation and removal of residual solvent molecules from samples. Dramatic improvement of crystallinity of as-grown samples was also achieved by methanol-soaking and then drying in air. Raman, infrared and X-ray diffraction results suggest that the methanol-soaked samples exhibit a solvated tetragonal structure. A significant finding is that the improved crystallinity by methanol-soaking did not degrade after removal of methanol molecules from samples by annealing. We examined the photo-polymerization of the structurally-improved C_{60} nanotubes, and found that the C_{60} nanotubes soaked into methanol were easier to polymerize than non-soaked samples.

Acknowledgments
We are grateful to K Nakama and T Yoshii at the Nippon Steel Glass Co., Ltd. for many useful discussions.

References

[1] Meng R L, Ramirez D, Jiang X, Chow P C, Diaz C, Matsuishi K, Moss S C, Hor P H and Chu C W 1991 *Appl. Phys. Lett.* **59** 3402
[2] Yosida Y 1992 *Japan. J. Appl. Phys.* **31** L505
[3] Liu J Z, Dykes J W, Lan M D, Klavins P, Shelton R and Olmstead M M 1993 *Appl. Phys. Lett.* **62** 531
[4] Miyazawa K, Obayashi and Kuwabara M 2001 *J. Am. Ceram. Soc.* **84** 3037
[5] Miyazawa K, Kuwasaki Y, Hamamoto K, Nagata S, Obayashi A and Kuwabara M 2003 *Surf. Interface Anal.* **35** 117
[6] Minato J, Miyazawa K and Suga T 2005 *Science and Technology of Advanced Materials* **6** 272
[7] Miyazawa K, Minato J, Yoshii T, Fujino M and Suga T 2005 *J. Mater. Res.* **20** 688
[8] Kobayashi K, Tachibana M and Kojima K 2005 *J. Cryst. Growth* **274** 617
[9] Fleming R M, Kortan A R, Hessen B, Siegrist T, Thiel F A, Marsh P, Haddon R C, Tycko R, Dabbagh G, Kaplan M L and Mujsce A M 1991 *Phys. Rev.* B **44** 888
[10] Minato J and Miyazawa K 2006 *Diamond & Related Materials* **15** 1151

Journal of Physics: Conference Series **159** (2009) 012021 doi:10.1088/1742-6596/159/1/012021

Synthesis and growth investigation of C_{60} fullerene nanowhiskers

K Hotta and K Miyazawa
Fullerene Engineering Group, Advanced Nano Materials Laboratory, National Institute for Materials Science, Tsukuba, Ibaraki, 305-0044, Japan

E-mail: HOTTA.kayoko@nims.go.jp

Abstract. The growth mechanism of C_{60} nanowhiskers (C_{60}NWs) is investigated. The C_{60}NWs were prepared by the liquid-liquid interfacial precipitation method (LLIP method). The length growth of C_{60}NWs was investigated for various volume ratios between C_{60}-saturated toluene solution and isopropyl alcohol. The length growth of C_{60}NWs varied depending on the volume ratio. The optimum volume ratio in preparing C_{60}NWs with a high uniformity was found to be 1:1. The water contained in isopropyl alcohol markedly influences the length growth of C_{60}NWs. C_{60}NWs did not form when isopropyl alcohol containing more than 2.5 mass % water was used.

1. Introduction

C_{60} was discovered by Kroto et al. in 1985 [1], and its properties and application in semiconductor, superconductor, medicine, composite materials and so forth have been studied. In 2001, C_{60} nanowhiskers (C_{60}NWs) were discovered [2] and the liquid-liquid interfacial precipitation method (LLIP method), i.e., the fabrication method of C_{60}NWs, was reported [3]. In the LLIP method, a poor solvent of fullerene is added to a good solvent solution of fullerene to make a liquid-liquid interface where fine fullerene crystals are formed and grow into fullerene nanowhiskers (FNWs). Using this method, the nanowhiskers of C_{60}, C_{70} and $C_{60}[C(COOC_2H_5)_2]$ [3, 4, 5], fullerene nanotubes (FNTs), i.e., tubular FNWs and fullerene nanosheets can be synthesized [6, 7, 8]. C_{60}NWs have a quasi one-dimensional single crystalline morphology with lengths of a few micrometers ~ several millimeters, diameters of a few ten nanometers ~ several hundred nanometers, a growth axis parallel to the [110] close-packed direction of fcc crystal system in a dried condition [3], and Young's modulus of about 32 ~ 54 GPa [9]. Tachibana et al. and Ringor et al. reported about photo-assisted growth of C_{60}NWs and C_{60} molecules were combined by van del Waals forces [10, 11, 12]. However, it is not clear about the growth mechanism of C_{60}NWs. To understand the growth mechanism of C_{60}NWs in the LLIP method is important for their future uses such as catalysts, templates for chemical synthesis, field-effect transistors and so forth [13, 14, 15]. Hence, in this study, the length growth of C_{60}NWs in various conditions is investigated to know the growth mechanism by using optical microscopy and scanning electron microscopy (SEM).

2. Experimental

In preparing C_{60}NWs by the LLIP method [3], toluene (99.5 % purity, 0.867 g/ml density, as-received non-dehydrated grade, Wako Pure Chemical Industries, Ltd., Japan) was used as a good solvent, and isopropyl alcohol (IPA) (99.7 % purity, 0.786 g/ml density, as-received non-dehydrated grade, Wako Pure Chemical Industries, Ltd., Japan) was used as a poor solvent of C_{60}. C_{60} powder (99.5 % purity, MTR Ltd., USA) were dissolved in toluene, ultrasonicated for 30 minutes and filtered to remove undissolved C_{60} powder. The C_{60}-saturated toluene solution was poured into a 10ml transparent glass bottle and IPA was slowly added to the C_{60}-saturated toluene solution to form a liquid-liquid interface in a water bath at 20 °C. Then, the solutions of the bottles were manually mixed and kept in an incubator (SANYO MIR-153) at 20 °C.

The C_{60}NWs were synthesized at various volume ratios of C_{60}-saturated toluene solution and IPA. The volume ratio of C_{60}-saturated toluene solution and IPA was set to be 1:9, 1:7, 1:5, 1:3, 1:1, 3:1, 5:1, 7:1, and 9:1. To investigate the effect of water contained in IPA, distilled water (0.998 g/ml density, Wako Pure Chemical Industries, Ltd., Japan) was added to IPA. The volume ratio of C_{60}-saturated toluene solution and IPA was 1:1. The water content in IPA was set to be 0, 0.4, 0.6, 0.9, 1.3, 2.5, 3.8, 5.0, 6.3, 7.5, 8.7, 9.9, 11.2 and 12.4 mass %. 1 day and 7 days after the start of synthesis of C_{60}NWs, the length of C_{60}NWs was measured by using an optical microscope (Nikon ECLIPSE ME600). The microstructure of C_{60}NWs was observed by SEM (JEOL JSM-6700F).

3. Results and Discussion

3.1. Effect of different solvent ratio on the synthesis of C_{60}NWs

C_{60}NWs could be synthesized at each solvent volume ratio of 1:9, 1:7, 1:5, 1:3 and 1:1, but the length of C_{60}NWs varied depending on the volume ratio. The length of C_{60}NWs synthesized by using the solution of 1:9 volume ratio was very short and many small particles were formed (Figure 1(a)). When the ratio was in the range of 1:7 ~ 1:3, the length of C_{60}NWs was still short, C_{60}NWs with lengths less than 5 μm were formed (Figure 1(b, c, d)). When the same amount of C_{60}-saturated toluene solution and IPA was mixed, the precipitates were composed of C_{60}NWs and granular crystals were not observed (Figure 1(e)). 2 types of crystal nuclei must have been formed. At first, nuclei are formed at the liquid-liquid interface, and after mixing the solution, more nuclei appear. Each nucleus grows in different ways depending on the solvent ratio of C_{60}-saturated toluene solution and IPA. With respect to the solutions with 3:1 ~ 9:1 ratios, C_{60}NWs were not formed. However, granular C_{60} crystals precipitated (Figure 1(f)). More uniform and long C_{60}NWs could be synthesized in the case of 1:1 solvent ratio. Ringor et al. reported that thinner and more uniform C_{60} nanotubes (C_{60}NTs), which are tubular C_{60} nanowhiskers, can be prepared using C_{60}-saturated pyridine solutions and IPA with solvent rations of 1:10 and 1:9 [12], though the solvent ratios to obtain high quality C_{60}NTs were different from our case. The solubility of C_{60} in toluene is 2.8 mg/ml and the solubility of C_{60} in pyridine is 0.89 mg/ml [16]. The solubility of C_{60} and the solvent ratio of IPA to C_{60}-saturated solution are found to influence the length growth of C_{60}NWs.

3.2. Effect of water content in IPA on the length growth of C_{60}NWs

C_{60}NWs could precipitate when the IPAs containing 0 ~ 1.3 % water were used in the LLIP method. The water content in the as-received IPA was measured using a Karl Fischer titrator (Coulometer 831 KF) before experiment. The initial water content in IPA was 0.017 %. When IPA containing 2.5 % water was used, C_{60}NWs with lengths greater than 10 μm could be synthesized in 1 day. But after 7 days from the start of synthesis, the C_{60}NWs were found to be partly decomposed and many bulky and granular C_{60} crystals appeared (Figure 2). When IPA containing more than 3.8 % water was used, no C_{60}NWs were formed, although granular C_{60} crystals precipitated. As shown in Figure 3, the length of C_{60}NWs increased with increasing the water content in IPA up to 2.5 mass % water. The chemical properties and cluster structure of IPA must be changed by the addition of water. A strong dipole - dipole interaction between IPA molecules and water molecules is conjectured to influence the length growth and morphology of C_{60}NWs.

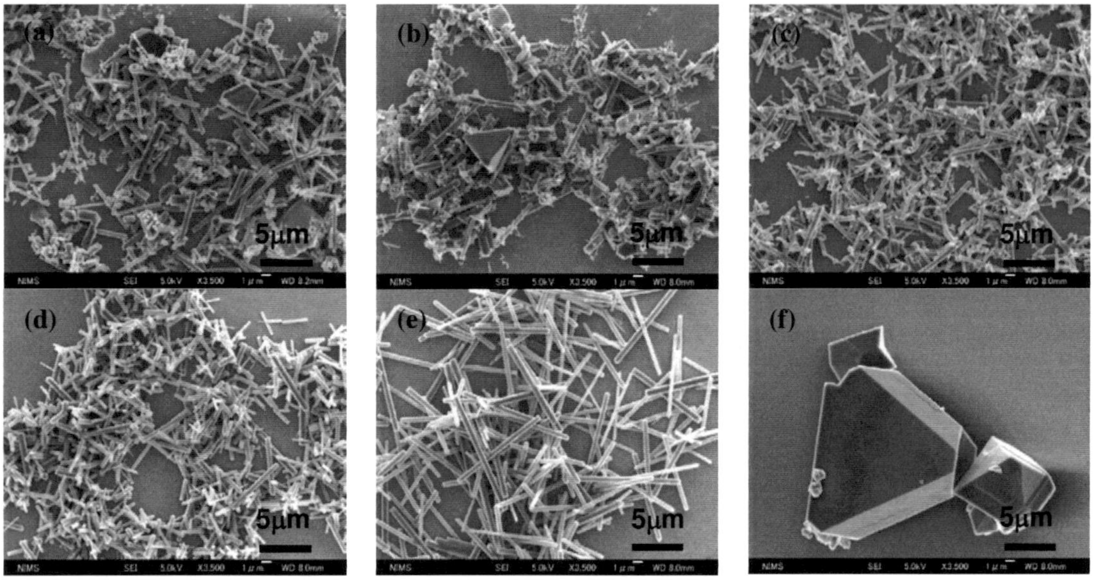

Figure 1. SEM images of C_{60}NWs. The solvent ratio of C_{60}-saturated toluene solution and IPA is (a) 1:9, (b) 1:7, (c) 1.5, (d) 1:3, (e) 1:1 and (f) 3:1, respectively.

Figure 2. Optical micrographs of C_{60}NWs prepared by using IPA containing 2.5 % water. Growth period of 1 day (a) and that of 7 days (b), respectively.

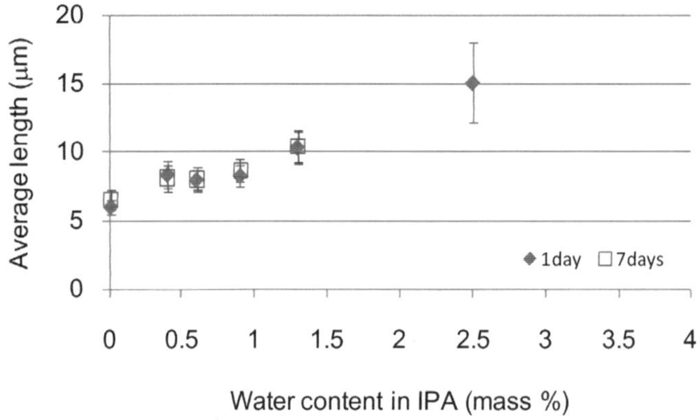

Figure 3. Average length of C_{60}NWs for the growth periods of 1 day and 7 days.

4. Conclusions

C_{60}NWs were synthesized using the solutions whose solvent ratio of C_{60}-saturated toluene solution and IPA was 1:9 ~ 1:1. In the range of 1:9 ~ 1:1 mixture ratios, C_{60}NWs were formed, while the C_{60}NWs with the highest uniformity were synthesized in the mixture of 1:1 ratio. However, C_{60}NWs could not be formed in the mixtures of 3:1 ~ 9:1. The length of C_{60}NWs increased with increasing the water content of IPA in the range of 0 ~ 1.3 mass % water. But when IPA with water content higher than 2.5 mass % was used, C_{60}NWs could not be synthesized and bulk C_{60} crystals precipitated. Small amount of water positively or negatively influences the length growth of C_{60}NWs. In future, the role of water to the growth of C_{60}NWs will be examined in comparison with the other polar solvents.

References

[1] Kroto H W, Heath J R, Brien S C O', Curl R F and Smalley R E 1985 *Nature* **318** 162
[2] Miyazawa K, Obayashi A and Kuwabara M 2001 *J. Am. Ceram. Soc.* **84** 3037
[3] Miyazawa K, Kuwasaki Y, Obayashi A and Kuwabara M 2002 *J. Mater. Res.* **17** 83
[4] Miyazawa K 2002 *J. Am. Ceram. Soc.* **85** 1297
[5] Miyazawa K, Mashino T and Suga T 2003 *J. Mater. Res.* **18** 2730
[6] Miyazawa K, Minato J, Yoshii T, Fujino M and Suga T 2005 *J. Mater. Res.* **20** 688
[7] Minato J and Miyazawa K 2006 *J. Mater. Res.* **21** 529
[8] Sathish M and Miyazawa K 2007 *J. Am. Chem. Soc.* **129** 13816
[9] Asaka K, Kato R, Miyazawa K and Kizuka T 2006 *Appl. Phys. Lett.* **89** 071912
[10] Tachibana M, Kobayashi K, UchidaT, Kojima K, Tanimura M and Miyazawa K 2003 *Chem. Phys. Lett.* **374** 279
[11] Kobayashi K, Tachibana M and Kojima K 2005 *J. Cryst. Grow.* **274** 617
[12] Ringor C L and Miyazawa K 2008 *Diam. Rel. Mat.* **17** 529
[13] Sathish M, Miyazawa K and Sasaki T 2007 *Chem. Mater.* **19** 2398
[14] Minato J and Miyazawa K 2006 *J. Mater. Res.* **21** 529
[15] Ogawa K, Kato T, Ikegami A, Tsuji H, Aoki N and Ochiai Y 2006 *Appl. Phys. Lett.* **88** 112109
[16] Ruoff R S, Tse D S, Malhotra R and Lorents D C 1993 *J. Phys. Chem.* **97** 3379

Journal of Physics: Conference Series **159** (2009) 012022 doi:10.1088/1742-6596/159/1/012022

Molecular dynamics and structural phase transition in C_{60} nanowhiskers

Hideaki Kitazawa, Kenjiro Hashi, Tuerxun Wuernisha, Kayoko Hotta, Cherry L Ringor, Takao Furubayashi, Atsushi Goto, Tadashi Shimizu, and Kun'ichi Miyazawa

National Institute for Materials Science, Tsukuba, Ibaraki, 305-0047, Japan

E-mail: KITAZAWA.Hideaki@nims.go.jp

Abstract. Nanowhiskers of C_{60} (C_{60} NWs) have been formed by a liquid-liquid interfacial precipitation method which uses an interface of the concentrated toluene solution of C_{60}/2-propanol. In order to study the motion of C_{60} molecules in nanowhiskers, we have performed measurements of low-temperature X-ray diffraction, magnetic susceptibility, specific heat and ^{13}C-NMR for C_{60} NWs. The temperature dependence of lattice constant for C_{60} NWs is in good agreement with that for pristine C_{60} powder, which demonstrates the large discontinuity at T_c ~ 265 K caused by the structural phase transition. However, the temperature dependence of magnetic susceptibility for C_{60} NWs exhibits a faint anomaly at T_c. The temperature dependence of specific heat for C_{60} NWs shows two small anomalies at 232 K and 254 K. High resolution ^{13}C-NMR measurements of the C_{60} NWs indicate that both of toluene and 2-propanol molecules interact with the C_{60} NW molecules in the suspension and the precipitate states obtained from the LLIP method. The broader linewidth of ^{13}C-NMR in the dried C_{60} NWs compared with that in the pristine C_{60} powder indicates that the slower molecular reorientation in the dried C_{60} NWs compared with that in the pristine C_{60} powder. Since a rapid rotation of C_{60} molecules could be directly detected by ^{13}C-NMR, polymer formation between C_{60} molecules is unlikely in C_{60} NWs at room temperature.

1. Introduction

Fullerene C_{60} nanowhiskers (C_{60} NWs) have been easily formed by a liquid-liquid interfacial precipitation (LLIP) method [1]. When 2-propanol (isopropyl alcohol) is gently added to a toluene solution saturated with C_{60}, fibrous fullerene single crystals with diameters as small as submicrons were precipitated at the interface between solutions. The initial TEM experiments suggested the existence of polymerization between C_{60} molecules along the growth axis [1]. However, subsequent X-ray diffraction (XRD) measurements revealed that the first precipitated crystals from the liquid phase have solvated structures with a hexagonal structure [2]. After the evaporation of solvent molecules in air, the structure rapidly changed into a face centered cubic (fcc) structure like a C_{60} solid. Since the Raman spectroscopic study revealed the photochemical polymerization [3], the irradiation of electron beam may affect the change of interaction between C_{60} molecules. Thus there is still uncertainty for the crystal structure of C_{60} NWs under the various experimental conditions. It is well known that the solid C_{60} undergoes an orientational ordering transition at T_c ~ 260 K. The phase transition is from nearly free rotating state of the molecules forming the fcc structure (space group

Fm3m) to a phase wherein the molecules undergoes only a ratchet type of correlated rotation about <111> directions of simple cubic (sc) structure (space group $Pa\bar{3}$) [4, 5]. If the polymerization between C_{60} molecules occurs in C_{60} NWs, some dynamics such as a rotation of C_{60} molecules observed above T_c and structural phase transition at T_c should be affected in C_{60} NWs. Since the ordering transition has been investigated with a wide range of techniques such as X-ray [4], neutron diffraction [6], NMR [5], and specific heat [7] etc., it is worth while understanding the dynamics of C_{60} molecules in C_{60} NWs by means of several kinds of techniques. We have performed measurements of low-temperature XRD, magnetic susceptibility, specific heat and ^{13}C-NMR for C_{60} NWs to obtain the information of C_{60} dynamics in C_{60} NWs.

2. Experimental procedures

The C_{60} NWs (dried C_{60} NWs in air) were prepared by the above mentioned LLIP method from pristine C_{60} powder (99.5%, MTR Ltd.). In order to determine the temperature dependence of lattice constants the powder XRD experiments were performed by using a Rigaku diffractometer (CuKα radiation) with a low-temperature option. The powder samples were mixed with a standard Si material (NIST SRM 640b) to obtain the more precise lattice constants. All the peaks below $2\theta = 35°$ were well indexed by the fcc crystal symmetry of the solid C_{60} except for small unknown peaks which may come from the remaining phases with solvated structure or other foreign phases. Each lattice constant *a* of C_{60} NWs and pristine C_{60} powder was estimated from the some assigned peaks below $2\theta = 35°$. The DC magnetic susceptibility $\chi(T)$ was measured in a magnetic field of 0.1T and 3 T for temperatures from 1.9 to 300 K by using a SQUID magnetometer (Quantum Design Co., Ltd.). The specific heat measurements were performed between 4 K and room temperature for pressed C_{60} NWs and pristine C_{60} powder with the relaxation method by a PPMS instrument (Quantum Design Co., Ltd). In order to investigate the interaction between solvent and C_{60} NW, NMR measurements were performed at 11.7 T (125.8 MHz for ^{13}C) using a JEOL ECA500 spectrometer. A 4 mm sample tube was used for a magic-angle-spinning (MAS) NMR measurement. Fourier transformed (FT) NMR spectra were obtained from free-induction-decay (FID) signals following a $\pi/2$ pulse of 2.4 μs. All NMR experiments were performed at room temperature. We have prepared three C_{60} NW samples for NMR experiments: suspension C_{60} NWs obtained by the LLIP method [1]; precipitate C_{60} NWs; and dried C_{60} NWs. The suspension C_{60} NW sample was packed in a glass tube with 4 mm diameter for NMR measurements. The precipitate C_{60} NWs were obtained from the suspension C_{60} NWs by centrifugation. Both of the suspension and the precipitate C_{60} NWs contain toluene and 2-propanol as solvent, while the dried C_{60} NWs contain no solvent.

3. Results and discussion

3.1. Low-temperature XRD

Figure 1 shows XRD patterns near the 220 peak of C_{60} NWs as a function of temperature. The peak position for the 220 reflection shifts largely between 260 K and 270 K. Figure 2 shows the temperature dependence of lattice constants for C_{60} NWs and pristine C_{60} powder. Both curves demonstrate large discontinuities at ca. 265 K which correspond to the structural phase transition at $T_c \sim 260$ K from the sc structure at low temperatures to the fcc structure at high temperatures [4]. Both curves are almost the same within the present experimental error bar. Overall temperature evolutions of lattice constants for both samples are in good agreement with the result determined from neutron diffraction [6] except for the anomaly at $T_{sg} \sim 90$ K which corresponds to the frozen rotation motion.

3.2. Magnetic susceptibility

Figure 3 shows the temperature dependence of magnetic susceptibility $\chi(T)$ for C_{60} NWs and pristine C_{60} powder. A clear step-like transition at 260 K was observed in the pristine C_{60} powder as shown in

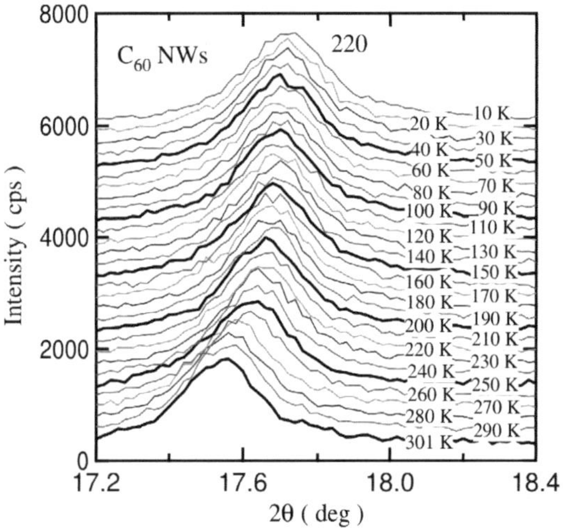

Figure 1. 220 peak in the XRD pattern of C_{60} NWs as a function of temperature. Each pattern is vertically shifted by +200 cps without overlapping.

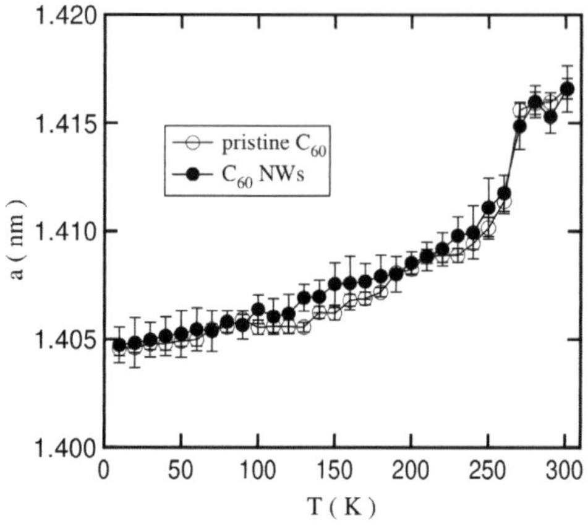

Figure 2. Temperature dependence of lattice constant a for C_{60} NWs and pristine C_{60} powder.

the inset of figure 3. This susceptibility discontinuity is attributed to the change in the intramolecular geometry at orientational order-disorder transition [8]. On the other hand, the discontinuity is very tiny but visible in the C_{60} NWs. This reduction of anomaly at 260 K suggests that the structural phase transition is incomplete or short-range in the C_{60} NWs below T_c. The overall field-cooled $\chi(T)$ curves of the C_{60} NWs and pristine C_{60} powder can be well fitted by the Curie-Weiss law $\chi(T) = C/(T-\theta_p) + \chi_0$ by the solid curves in figure 3. The C, θ_p and χ_0 is the Curie constant, $Np_{\mathrm{eff}}^2/3k_B$, the paramagnetic Curie temperature, and temperature-independent term, respectively, where N, p_{eff}, and k_B is a number of magnetic moments, the effective moment, and the Boltzmann constant. If the spin amplitude of magnetic moment is assumed to be 1/2, the number of magnetic moments corresponds to 1.20 % and 0.693 % in a unit cell for C_{60} NWs and pristine C_{60} powder, respectively. The estimated magnetic moments may originate from any magnetic impurity. The obtained parameters are listed in Table 1. The $\chi(T)$ curves of pristine C_{60} powder below ca. 100 K depends on between zero-field-cooled (ZFC) and field-cooled (FC) processes, but the $\chi(T)$ curve of C_{60} NWs does not exhibit such a hysteresis. The hysteresis can be explained in terms of a transition into a frozen magnetic glass state below T_{sg} ~ 90 K [9].

Table 1. Magnetic parameters of C_{60} NWs and pristine C_{60} powder.

	m(mg)	χ_0(emu/mol)	C(emu/molK)	$\theta_p(K)$	N(/mol)	N(%/unit cell)
C_{60} NWs	27.3	-1.521(6) $\times 10^{-3}$	4.51(2) $\times 10^{-3}$	-1.61(2)	7.24 $\times 10^{21}$	1.20
pristine C_{60}	38.0	-1.154(4) $\times 10^{-3}$	2.60(2) $\times 10^{-3}$	-1.61(2)	4.17 $\times 10^{21}$	0.693

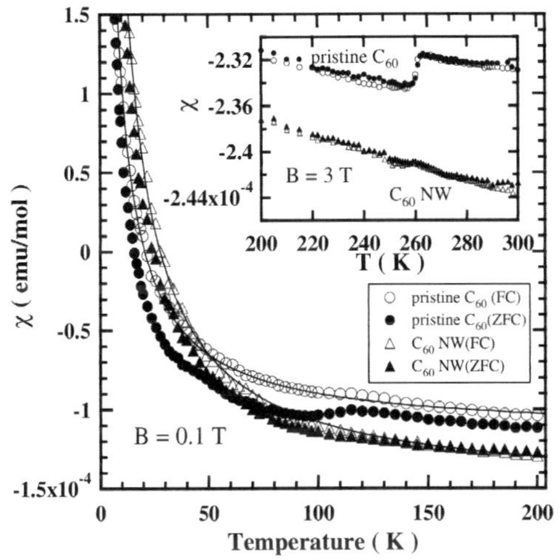

Figure 3. Temperature dependence of magnetic susceptibility for pressed C_{60} NWs and pristine C_{60} powder. The solid lines represent the fitting curves by the Curie-Weiss law. The inset shows the precise data near 260K.

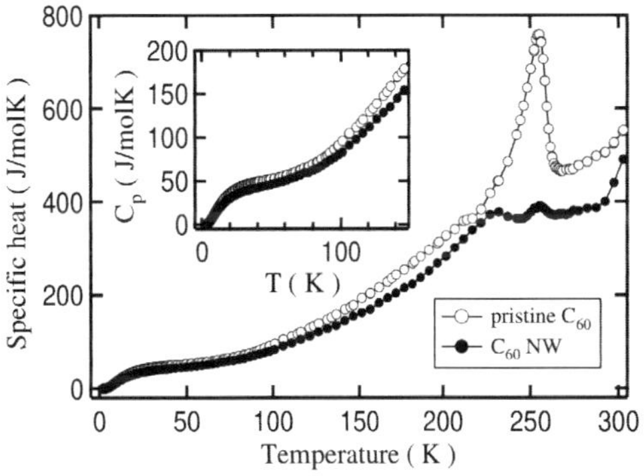

Figure 4. Temperature dependence of specific heat for pressed C_{60} NWs and pristine C_{60} powder.

3.3. Specific heat

Figure 4 shows the temperature dependence of specific heat for pressed C_{60} NWs and pristine C_{60} powder. A large peak at 254 K in the pressed pristine C_{60} powder must be caused by this structural phase transition [7]. The single anomaly observed in the pressed pristine C_{60} splits into two small anomalies in the pressed C_{60} NWs at 254 K and 232 K. The former anomaly at 254 K is smaller than that for pressed pristine C_{60} powder. The latter anomaly at 232 K is not observed in the pressed pristine C_{60} powder. This splitting of anomaly at 254 K in the pressed pristine C_{60} powder is a strong evidence of suppression of the structural phase transition in C_{60} NWs. The specific heat in pressed C_{60} NWs starts to deviate from 6 K to higher temperature. This means that the entropy in pressed C_{60} NWs is reduced up to room temperature as compared with that in pressed pristine C_{60} powder.

3.4. NMR

Figure 5 shows the NMR spectra for the suspension of C_{60} NWs with and without a proton decoupling. The proton decoupling was performed by the WALTZ pulse sequence. A peak at 143 ppm comes from the C_{60}NWs and the remaining peaks come from toluene. The splits of the toluene peaks due to the *J* coupling with protons were eliminated by the proton decoupling. It is interesting that the peak of C_{60}NWs is also affected by the proton decoupling. This indicates that the protons of the solvent exist near the C_{60} NW and the C_{60} NW interacts with the proton of the solvent. It seems that the linewidth of the C_{60} NW peak without the proton decoupling, originates from dipole fields of the proton, because MAS-NMR spectrum of C_{60} NW (not shown) has a narrow single peak. The narrowing of the linewidth of C_{60} NW peak by the proton decoupling is caused by partial decoupling of the dipole field of the proton with the WALTZ pulse sequence.

To clarify the effect of the interaction between the C_{60} NW and solvent on NMR spectra, NMR measurements were performed on the precipitate sample. Figure 6 shows the proton decoupled NMR spectra of the suspension and the precipitate C_{60} NWs. All peaks of toluene and 2-propanol in the

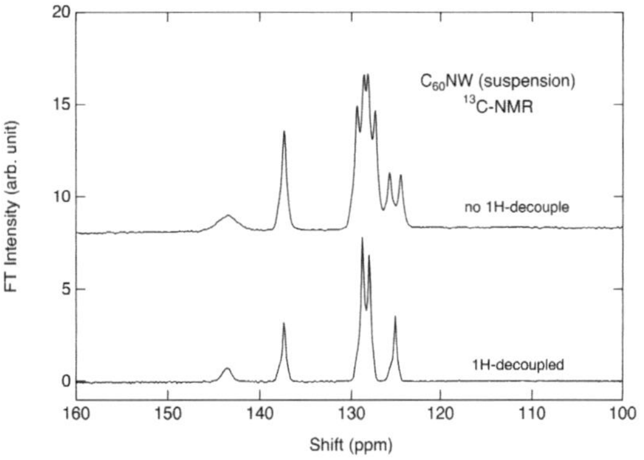

Figure 5. NMR spectra of the suspension of C_{60} NWs at 11.7 T with and without the proton decoupling. The proton decoupling was performed by the WALTZ pulse sequence.

precipitate sample were broadened compared with the corresponding peaks of the suspension sample. This means that both of toluene and 2-propanol molecules cannot reorient freely in the precipitate state. It seems that both of toluene and 2-propanol molecules are bound by C_{60} NW in the precipitate sample.

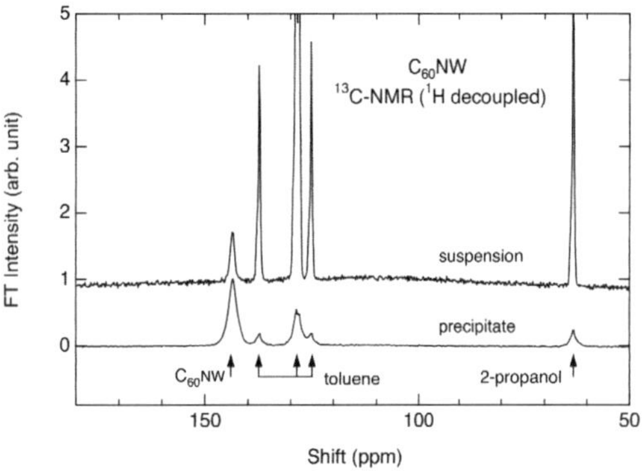

Figure 6. Proton decoupled NMR spectra of the suspension and the precipitate C_{60} NWs at 11.7 T. The proton decoupling was performed by the WALTZ pulse sequence.

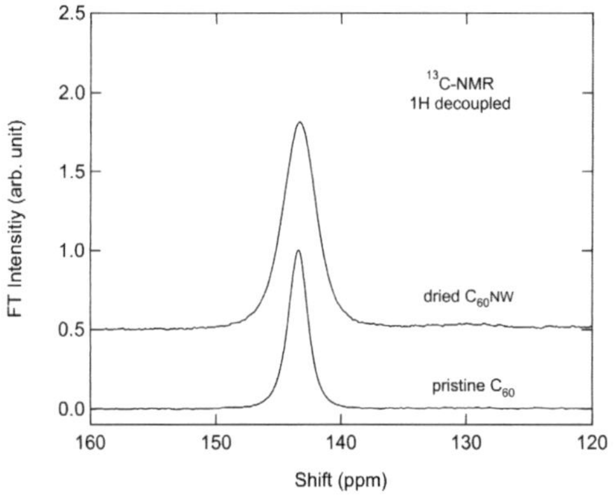

Figure 7. Proton decoupled NMR spectra of the dried C_{60} NWs and pristine C_{60} powder at 11.7 T. No signal of solvent was observed for the dried C_{60} NWs.

The peak of the C_{60} NWs in the precipitate sample is also broader than that in the suspension sample. This indicates that the dipole field from the protons of the solvent in the precipitate sample is larger than that in the suspension sample. It seems that, however, there is another reason for the broadening. The NMR spectra of the dried C_{60} NWs and pristine C_{60} powder show a single peak at 143 ppm as shown in figure 7. The linewidth of the dried C_{60} NWs is larger than that of pristine C_{60} powder. Both of the linewidth of the dried C_{60} NWs and the pristine C_{60} powder are not affected by the proton decoupling, because there is no solvent in the samples. It seems that the difference of the linewidth between the dried C_{60} NWs and the pristine C_{60} powder comes from the difference in the molecular dynamics. Rapid and isotropical reorientation of the pristine C_{60} molecules averages out the chemical shift anisotropy (CSA) at room temperature [5]. The CSA powder pattern is recovered gradually with decreasing temperature, because the molecular reorientation becomes slower with decreasing temperature. It seems that a slower molecular reorientation of the dried C_{60} NWs is responsible for the broader linewidth of the dried C_{60} NWs compared with the pristine C_{60} powder. The difference of the molecular dynamics between the dried C_{60} NWs and the pristine C_{60} powder also affects the spin-lattice relaxation time T_1. The T_1 for the dried C_{60} NWs is about 18 s, while that for the pristine C_{60} powder is 38 s. Since a rapid rotation of C_{60} molecules could be directly detected by ^{13}C-NMR, polymer formation between C_{60} molecules is unlikely at room temperature. Measurements of the temperature dependence of the relaxation time T_1 for the dried C_{60} NWs is required for more detailed discussion of the molecular dynamics.

4. Conclusion

The present research can be summarized as follows. The low-temperature XRD of C_{60} NWs measurements revealed the existence of the structural phase transition at T_c ~ 255 K from the sc structure at low temperatures to the fcc structure. However, the strong suppression of anomalies in the magnetic susceptibility and the specific heat of C_{60} NWs at T_c suggest the above-mentioned phase transition is incomplete or short-range. High resolution ^{13}C-NMR measurements of the C_{60} NWs indicate that both of toluene and 2-propanol molecules interact with the C_{60} NW molecules in the suspension and the precipitate states obtained from the LLIP method. The broader linewidth of the dried C_{60} NW compared with that of the pristine C_{60} powder indicates that the slower molecular reorientation in the dried C_{60} NW compared with that in the pristine C_{60} powder. Thus, judging from the degree of peak width in C_{60} NW, the formation of polymerization must be not realized in room temperature. The polymerization between C_{60} molecules along the growth axis suggested by the TEM observation [1] may be the consequence of the electron beam irradiation.

Acknowledgments

One of the authors K.H. appreciates Dr. M. Murakami and Dr. K. Deguchi for technical support of and fruitful discussions. This work of H. K. is partly supported by Grant-in-Aid for Scientific Research on priority Areas "High Field Spin Science in 100T"(No.451) from the Ministry of Education, Culture, Sports, Science and Technology (MEXT).

References

[1] Miyazawa K, Kuwasaki Y, Obayasi A and Kuwabara M 2002 *J. Mater. Res.* **17** 83
[2] Minato J and Miyazawa K 2005 *Carbon* **43** 2837
[3] Tachibana M, Kobayashi K, Uchida T, Kojima K, Tanimura M and Miyazawa K 2003 *Chem. Phys. Lett.* **374** 279
[4] Heiney P A, Fischer J E, McGhie A R, Romanow W J, Denenstein A M, McCauley J P Jr, Smith A B III and Cox D E 1991 *Phys. Rev. Lett.* **66** 2911
[5] Tycko R, Dabbagh G, Fleming R M, Haddon R C, Makhija A V and Zahurak S M 1991 *Phys. Rev. Lett.* **67** 1886
[6] David W I F, Ibberson R M, Dennis T J S, Hare J P and Parssides K 1992 *Europhys. Lett.* **18** 219
[7] Matsuo T, Suga H, David W I F, Ibberson R M, Bernier P, Zahab A, Fabre C, Rassat A and Dworkin A 1992 *Solid State Comm.* **83** 711
[8] Luo W, Wang H, Ruoff R S, Cioslowski J and Phelps S 1994 *Phys. Rev. Lett.* **73** 186
[9] Buntar V, Weber H W and Ricco M 1995 *Solid State Comm.* **98** 175

Improved Fullerene Nanofiber Electrodes used in Direct Methanol Fuel Cells

Q Wang[1], Y Zhang[2], K Miyazawa[3], R Kato[3], K Hotta[3] and T Wakahara[3]

[1] Nano Craft Technologies Co., Ltd. , Tsukuba, Japan
[2] Nationals Institute of Advanced Industrial Science and Technology , Tsukuba, Japan
[3] National Institute for Materials Science, Tsukuba, Japan

E-mail: yi.zhang@aist.go.jp (Dr. Zhang), q.wang@aist.go.jp (Dr. Wang)

Abstract. Platinum supported on fullerene nanofibers as possible electrodes for direct methanol fuel cells (DMFC) were studied. Fullerene nanofiber with 20 wt% Pt loading was mixed with 5 wt% Nafion solution. The mixture paste was coated on Nafion 117 membrane and sandwiched with silicon plates. To increase the surface reaction area of catalyst, nanoimprint was used to fabricate micro-patterns in the Nafion proton exchange membrane. Nanoimprint pattern consisted of dots of 500 nm-in-diameter, 140 nm-in-depth and 1 μ m-in-spacing. The nanoimprint of the treated proton exchange membrane (PEM) was carried out in a desktop thermal nanoimprint system (NI273, Nano Craft Tech. Corp., Japan) at the optimized conditions of 130 °C and pressure of 3 MPa for 6 min. Then the Pt-coated PEM was sandwiched with micro-channelled silicon plates to form a micro-DMFC. With passively feeding of 1 M methanol solution and air at room temperature, the as-prepared cell had the open circuit voltage of 0.34 V and the maximum power density of 0.30 mW/cm^2. Compared with a fresh cell, the results shows that nanofibers used in nanoimprinted PEM have an improvement on the performance of micro fuel cells.

1. Introduction

Fuel cells are becoming an attractive and viable alternative to battery packs for portable electronic devices. In the different types of fuel cells, direct methanol fuel cells (DMFCs) have attracted the attention of many researchers, because of its characteristics as a small, lightweight and off-plug power generating system. However, how to improve the catalyst utility and efficiency is still the major bottle-neck in the development of DMFCs. Many efforts have been devoted to improve the catalyst utility and efficiency using MEMS-based methods. Feng et al. [1] recently developed micro-pillar structure and porous polypyrrole/Nafion® composite membrane for micro-fuel cell application. Yeom et al. [2] fabricated silicon meshes as the catalyst support and current collector for MEMS-based fuel cell application. Seo and Cho [3] succeeded in increasing the OCV from 0.2 to 0.85 V for their MEMS-based DMFC with silicon-based rugged electrode. However, it is found that most of MEMS-based DMFCs show much lower performance than the traditional DMFCs consisting of the graphite-based porous electrodes and proton exchange membrane (PEM) assemblies [3-6]. On the other hand, it is recognized that the form of catalyst support can affect the dispersion of the catalyst particles [7]. In the last several years, carbon nanotubes and nanofibers have been proposed as Pt supports for proton exchange membrane fuel cells and direct methanol fuel cells [8-9].

Journal of Physics: Conference Series **159** (2009) 012023 doi:10.1088/1742-6596/159/1/012023

In this work, Platinum supported on fullerene nanofibers as possible electrodes for direct methanol fuel cells was studied. We applied nanoimprint to fabricate micro-patterns in the Nafion proton exchange membrane with catalyst layers to increase the surface reaction area further. The nanoimprint pattern on the PEM could possibly facilitate the three-phase contact of fuels, catalyst and electrolyte in the three-dimensional zone by increasing the surface area and improving the flow field of reactants. Therefore, high performance seems possible for MEMS-based DMFC even with the absence of porous electrode.

2. Experimental

Fullerene nanofibers heat-treated at 900°C in vacuum with 20 wt% Pt loading were mixed with 5 wt% Nafion solution. The mixture paste was coated on Nafion 117 membrane and sandwiched with silicon plates. To increase the surface reaction area of catalyst, nanoimprint was used to fabricate micro-patterns in the Nafion proton exchange membrane. Nanoimprint pattern consisted of dots of 500 nm-in-diameter, 140 nm-in-depth and 1 μ m-in-spacing. Silicon was used as mold material. The dot pattern was prepared on silicon wafer by electron beam lithography. The nanoimprint of the treated PEM was carried out in a desktop thermal nanoimprint system (NI273, Nano Craft Tech. Corp., Japan) at the optimized conditions of 130 °C and pressure of 3 MPa for 6 min. The de-molding temperature was 60 °C. The nanoimprinted PEM was examined by atomic force microscope (AFM) (SPA501, SII, Seiko Instruments Inc.). Thick Pt layer (20 nm) was sputtered as catalyst layer directly on both sides of the nanoimprinted PEM at Ar pressure of 0.2 Pa and power of 100 W [10]. The 20 nm-thick Pt coated PEM was observed by scanning electron microscope (SEM) (JSM 5500, JEOL).

The Pt-coated PEM was mechanically sandwiched with the Pt-coated silicon plates using Teflon plate to form a micro-DMFC (Fresh Cell). To identify the effects of nanoimprint, we also prepared a micro-DMFC using Nafion® 117 membrane without the nanoimprint modification (Fullerene Cell). The as-prepared cells were tested with a potentiostat (SI1287 Electrochemical Interface, Solartron) at room temperature using passive fuel feeds. Aqueous 1 M methanol solution was fed into the cell through the through-hole with a syringe. When open circuit voltage (OCV) reached a stable value, the cell performance was characterized from current–voltage (I–V) curves at a scan rate of 10 mV/s.

3. Results and discussion

Fig. 1a shows SEM image of the as-prepared silicon mold. The dot was 500 nm-in-diameter and 1 μ m-in-spacing. Fig. 1b and c shows SEM images of the nanoimprinted Nafion® 117 membrane. Fig. 1d shows AFM image of fine pattern on the Nafion® 117 membrane formed by the nanoimprint procedure. Micro-convexes were obtained and uniformly distributed on the PEM. The micro-convexes were about 600 nm in diameter at the bottom. The height of the micro-convexes ranged from 50 to 70 nm. The size of the pattern on the PEM had deviated from that of the pattern on the mold. It was mainly due to the relaxation of patterns formed in polymers by the nanoimprint. There were many tiny cracks visible in the Pt layer on Nafion® 117 membrane (see Fig. 1c). The cracking of Pt layer was owing to that Nafion® 117 membrane had changed in volume with the degree of its humidification. This phenomenon was also found at prior observations [11-12]. Fig. 2 is an optical image of the DMFC using nanoimprinting treated PEM with micro-fabricated channels for flow field and current collector.

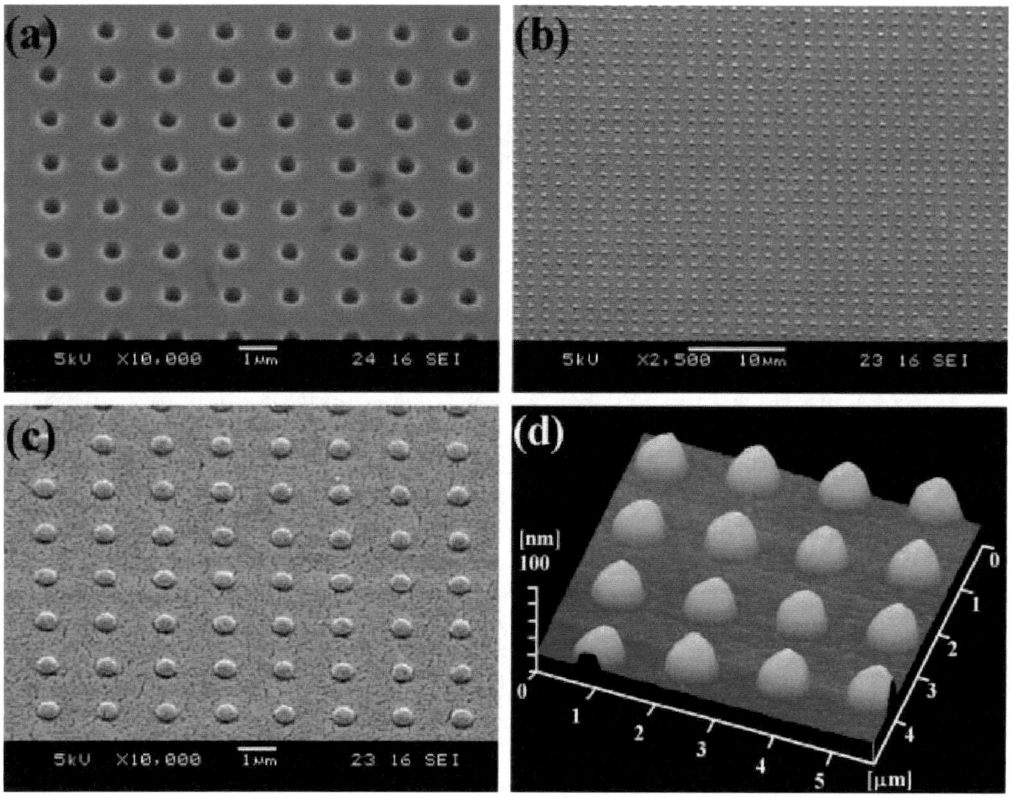

Fig.1 SEM images of as-prepared silicon mold (a), as-prepared fine pattern on the Nafion® 117 membrane (b and c) and corresponding AFM image (d).

Fig. 2 Optical image of the DMFC using nanoimprinting treated PEM

Fig. 3 shows measured performance of Fresh Cell and Fullerene Cell. The results of the Fresh Cell showed the OCV of 0.4 V and the maximum power density of 0.025 mW/cm^2. Although, the performance of the Fresh Cell was poor, it was similar to that of the state-of-the-art MEMS-based DMFC with planar electrode. Onoe et al. [13] reported that the OCV and the maximum power density of a MEMS-based DMFC were 0.15 V and 16 μ W/cm^2, respectively, with the feeding of 1 M methanol solution. Sim et al. [10] reported that a MEMS-based DMFC had the OCV of 0.1 V, but its power out was not available, because Au layer was used as catalyst. In addition, pure Pt catalyst was subject to the poisoning of CO in the Fresh Cell. For the Fullerene Cell, it showed the OCV of 0.33 V and the maximum power density of 0.03 mW/cm^2. The maximum power density of Fullerene Cell is

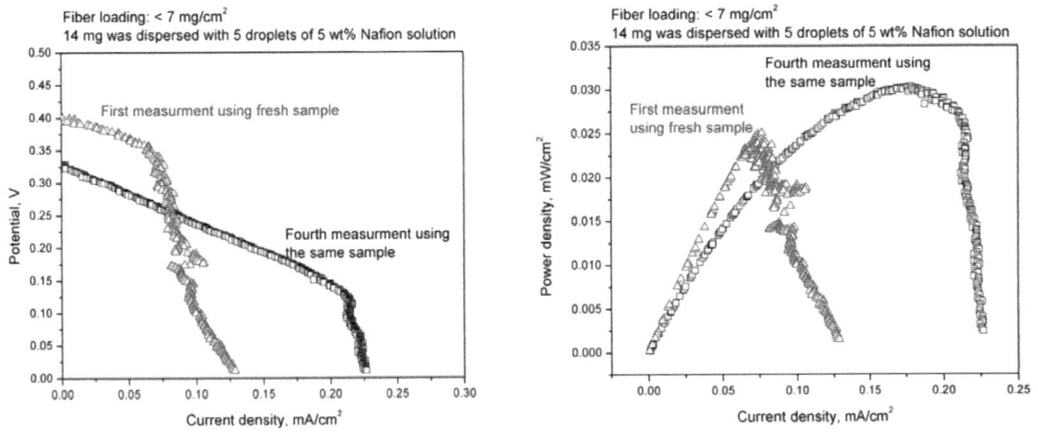

Fig.3 Measured performance of DMFC Cells at room temperature using passively 1 M methanol solution and air feeds.

higher than that of the Fresh Cell, although the OCV of Fullerene Cell is lower. Compared with previous researchers' studies, we found that the performance of Fullerene Cell was comparable to a micro-DMFC consisting of commercially graphite-based porous electrode and membrane assemblies [14]. The micro-DMFC in the literature [14] showed the OCV of 0.7 V and the maximum power density of 0.29 mW/cm^2 with passively feeding of 6 M methanol solution. This makes us believe that Fullerene Cell using the nanoimprinting treated PEM had a better performance compared with the Fresh Cell. We think that the performance improvement was mainly resulted from the factors as follows. Firstly, the apparent surface area of PEM increased by 12% with the nanoimprint modification. Secondly, the coverage of Pt film was poorer on the nanoimprinted PEM than that on the flat PEM. The Pt film in particular had much poorer coverage but more porous on the side wall of the micro-convex pattern because it was prepared by direct sputtering in this work. As a result, the practical surface area was increased further. Finally, the catalyst efficiency was also increased owing to that the catalyst support fullerene nanofibers which increased the surface reaction area on the surface of PEM.

From Fig. 3 we also found that the measured potential of the Fresh Cell reduced abruptly at the current of about 0.07 mA/cm^2, while the phenomenon was absent for the Fullerene Cell. This might be related to the adopted channel structure and the CO poisoning of Pt catalyst in this work. The channel structure was not perfect for the passively feeding of fuels and the purging of products such as CO_2, so that it was difficult to maintaining fuel concentrations near PEM as the initial value. The CO poisoning of Pt catalyst was also severe. This means that higher cell performance could be expected with the improvement of fuel feeds and the optimization of catalyst layer, as well as the optimization of the nanoimprint pattern.

4. Conclusions

In this work, Platinum supported on fullerene nanofibers as possible electrodes for direct methanol fuel cells was studied. Heat-treated fullerene nanofibers with 20 wt% Pt loading were mixed with 5 wt% Nafion solution. The mixture paste was coated on Nafion 117 membrane and sandwiched with silicon plates. To increase the surface reaction area of catalyst, nanoimprint was used to fabricate micro-patterns in the Nafion proton exchange membrane. Then the Pt-coated PEM was sandwiched with micro-channeled silicon plates to form a micro-DMFC. With passively feeding of 1 M methanol solution and air at room temperature, the as-prepared cell had the open circuit voltage of 0.34 V and the maximum power density of 0.30 mW/cm^2. The results were comparable to a micro-DMFC consisting of commercially graphite-based porous electrode and membrane assemblies. It shows that fullerene nanofibers used in nanoimprinted PEM have an improvement on the performance of micro direct methanol fuel cells.

References

[1] Feng C, Chan P C H and Hsing I-M 2007 *Electrochem. Commun.* **9** 893
[2] Yeom J, Mozsgai G Z, Flachsbart B R, Choban E R, Asthana A, Shannon M A and Kenis P J A 2005 *Sensor. Actuat.* B **107** 882
[3] Seo Y H and Cho Y H 2003 *Proc. IEEE MEMS'03* 375
[4] Nguyen N T and Chan S H 2006 *J. Micromech. Microeng.* **16** R1
[5] Wozniak K, Johansson D, Bring M, Velasco A S and Enoksson P 2004 *J. Micromech. Microeng.* **14** S59
[6] Ito T, Kimura K and Kunimatsu M 2006 *Electrochem. Commun.* **8** 973
[7] Dicks A L 2006 *J. Power Sources* **156** 128–141
[8] Carmo M, Paganin V A, Rosolen J M and Gonzalez E R 2005 *J. Power Sources* **142** 169
[9] Li W, Liang C, Zhou W, Qiu J, Zhou Z, Sun G and Xin Q 2003 *J. Phys. Chem.* B **107** 6292
[10] Yeom J, Mozsgai G Z, Flachsbart B R, Choban E R, Asthana A, Shannon M A and Kenis P J A 2005 *Sensor. Actuat.* B **107** 882
[11] Hayre R O, Lee S J, Cha S W and Prinz F B 2002 *J. Power Sources* **109** 483
[12] Gruber D, Ponath N, Müller J and Lindstaedt F 2005 *J. Power Sources* **150** 67
[13] Onoe S, Tanaka H, Hoshino K, Matusmoto K and Shimoyama I 2005 *Proc. Transducers's 05* 1296
[14] Sim W Y, Kim G Y and Yang S S 2001 *Proc. IEEE MEMS'01* 341
[15] Wozniak K, Johansson D, Bring M, Velasco A S and Enoksson P 2004 *J. Micromech. Microeng.* **14** S59

Journal of Physics: Conference Series **159** (2009) 012024 doi:10.1088/1742-6596/159/1/012024

High-resolution Transmission Electron Microscopy of Heat-treated C_{60} Nanotubes

R Kato[1], K Miyazawa[1], T Nishimura[1] and Z M Wang[2]

[1]*National institute for Materials Science, Tsukuba, Ibaraki, 305-0044, Japan*

[2]*Energy Technology Research Institute, AIST, Tsukuba, Ibaraki 305-8569, Japan*

E-mail: KATO.Ryoei@nims.go.jp

Abstract. Fullerene nanowhiskers and nanotubes are single crystals consisting of C_{60} molecules with a high aspect ratio of length to diameter. In this study, C_{60} nanowhiskers and nanotubes were heat-treated at various temperatures up to 3000 °C. The structural characteristics of the heated samples were examined using transmission electron microscopy (TEM) and Raman spectroscopy. By the high-temperature heat treatment, the C_{60} nanotubes turned to glassy carbon composed of randomly oriented ribbons with graphitic layers with an interlayer spacing which is close to that of the (002) planes of graphite.

1. Introduction

In 2001, Miyazawa et al. synthesized single crystalline non-tubular nanowhiskers consisting of C_{60} molecules with a high aspect ratio of length to diameter, typically a diameter of sub-micrometer and a length of more than 100 μm [1]. Furthermore, tubular C_{60} nanowhiskers, i.e., C_{60} nanotubes were fabricated by the same method as the C_{60} nanowhiskers [2]. The dried fullerene nanofibers have a face-centered cubic structure or a body-centered tetragonal structure [3], and the lattice constant of the C_{60} nanofibers with the face-centered cubic structure is close to that of pristine C_{60} crystal.

The as-prepared C_{60} whiskers have a high electrical resistivity $>10^5$ Ωcm, but it is reported that the resistivity decreases with decreasing diameter [3, 4].

On the other hand, with respect to the thermal properties of C_{60} nanowhiskers, the heated fullerene nanowhiskers could get a high specific surface area and a high electrical conductivity [5]. Asaka et al. examined the electrical conductivity of a C_{60} whisker preheated in vacuum at 1100 °C in a TEM [6]. The preheated C_{60} whisker was broken when a high current was applied. Interestingly, a whisker composed of multilayered carbon hollow nanocapsules less than 10 nm in diameter was produced. It was deduced that the formation of the capsules results from an increase in temperature due to the resistive Joule heating.

The whisker composed of carbon nanocapsules has a high specific surface area and a pours structure. Therefore the high temperature-treated fullerene nanowhiskers exhibit a high specific surface area and a good electrical conductivity. They are expected to be utilized as new electrode materials for fuel cells.

In this study, C_{60} nanowhiskers and C_{60} nanotubes, i.e., C_{60} nanofibers, were heat-treated at various temperatures up to 3000 °C. The structural characteristics of the samples heat-treated at each temperature were examined using optical microscopy, scanning electron microscopy (SEM), TEM and Raman spectroscopy.

2. Experimental

The fullerene nanowhsikers were prepared by the liquid-liquid interfacial precipitation method [7]. Firstly, C_{60} powder (99.95% pure C_{60}, MTR Ltd., USA) was dissolved in toluene by ultrasonication for 30 minutes. The toluene solution saturated with C_{60} was filtered by a paper filter to remove undissolved C_{60} powder. Isopropyl alcohol was gently added to the C_{60}-toluene solution to form the liquid-liquid interface in a glass bottle. The bottles were stored in an incubator for a few days.

The fullerene nanotubes were also fabricated by a similar liquid-liquid interfacial precipitation method [2, 8, 9], where a pyridine solution saturated with C_{60} irradiated by visible light (436 nm) or ultraviolet light (302 nm) was used.

The grown C_{60} nanofibers in a glass bottle were collected by centrifugal separation method or filtration. The prepared C_{60} nanofibers were heated for 60 min at 900 °C in an evacuated fused silica tube set in a muffle furnace (Denken Co., Ltd., KDF-S70). The C_{60} nanofibers heated at 900 °C were collected from the silica tube and further heated at 2000 °C, 2500 °C and 3000 °C, respectively, for one hour in Ar atmosphere using an ultra-high temperature furnace. The microstructure of samples was examined using a high-resolution transmission electron microscope (HRTEM, JEM-4010, JEM-2010, JEOL) and a microscopic Raman spectrophotometer (JASCO, NRS-3100, Japan) at a laser excitation wavelength of 532 nm.

3. Results and discussion

Figures 1 (a) and (b) show optical microscopy image of as-prepared C_{60} nanotubes on a slide glass and scanning electron microscopy image of that on a silicon plate, respectively. The C_{60} nanotubes exhibit metallic red brownish colour and grow to a length of more than 100 micrometer order and have a smooth surface.

Figure 2 shows the low magnification and high-resolution TEM images and electron diffraction patterns of the as-prepared or heat-treated C_{60} nanotubes that were dispersed on TEM microgrids. The C_{60} as-prepared nanotube has a single crystalline structure. But the C_{60} nanotube heated at 900 °C

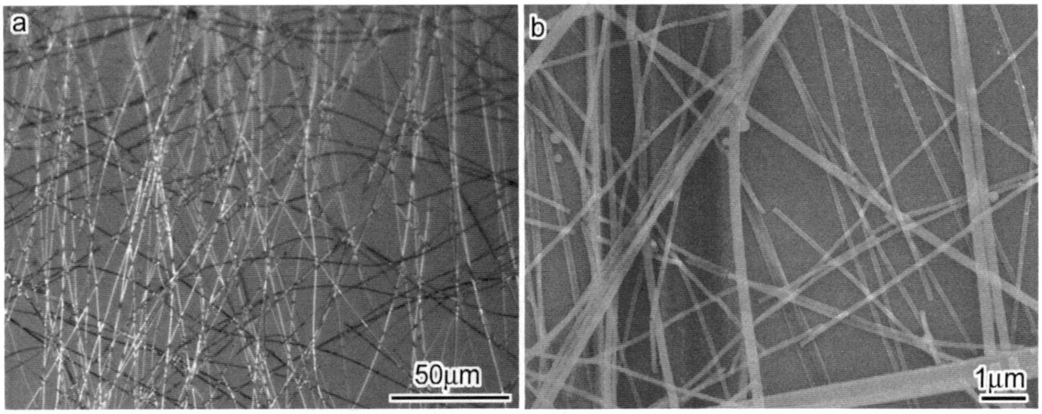

Figure 1. (a), (b) Optical microscopy image of as-prepared C_{60} nanotubes on a slide glass and scanning electron microscopy image of that on a silicon plate, respectively.

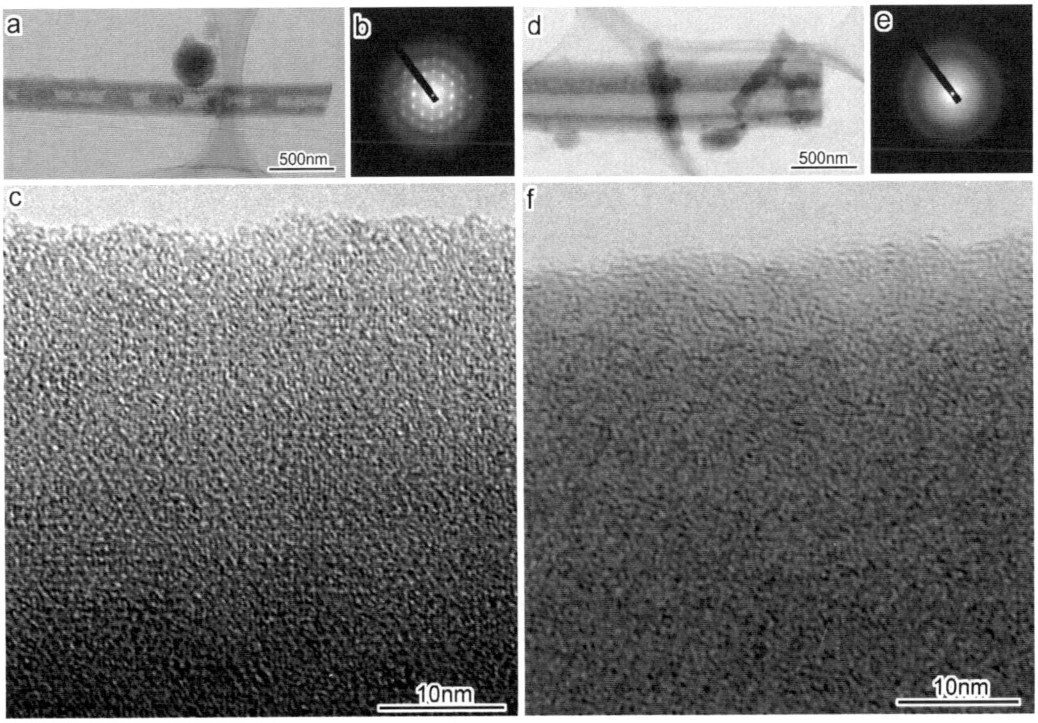

Figure 2. (a) Conventional TEM (CTEM) image, (b) electron diffraction pattern and (c) high-resolution TEM image of an as-prepared C_{60} nanotube. (d) CTEM image, (e) electron diffraction pattern and (f) high-resolution image of the C_{60} nanotube heated at 900 °C.

shows an amorphous structure as shown Figs. 2 (e) and (f).

Figure 3 shows the Raman spectra of heat-treated non-tubular C_{60} nanowhiskers, as-prepared C_{60} nanowhiskers and graphite. The Raman spectrum of the as-prepared C_{60} nanowhiskers dried in air at room temperature is very similar to that of pristine C_{60} [10]. Hence, those C_{60} nanowhiskers without heat treatment are found to be composed of the C_{60} molecules that are loosely bound via van der Waals bonding forces like the case of pristine face-centered cubic C_{60} crystals. The change of Raman spectra for the heat-treated C_{60} nanowhiskers is shown in Fig. 3. The peaks observed in the nanowhiskers without heat treatment are not found in the nanowhiskers heated at 900 °C. The peak at 1461cm^{-1} as shown in the spectrum of the as-prepared C_{60} nanowhisker shows the A_g (2) peak pentagonal pinch mode which is caused by the vibration of 5-membered rings of C_{60} cage [11]. The disappearance of this peak suggests that the

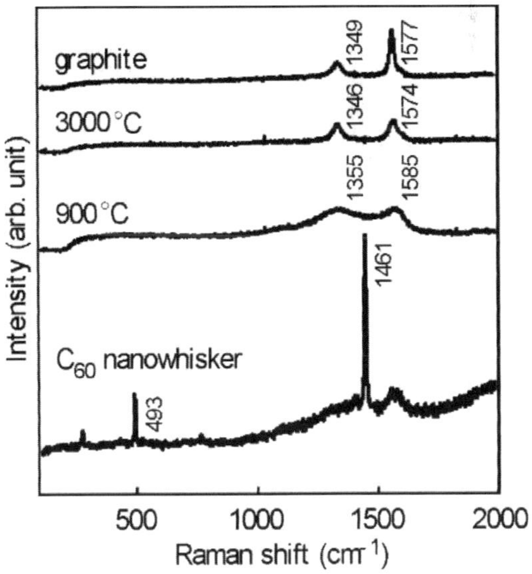

Figure 3. Raman spectra of as-prepared C_{60} nanowhiskers, the C_{60} nanowhiskers heated at 900°C and 3000 °C, and graphite.

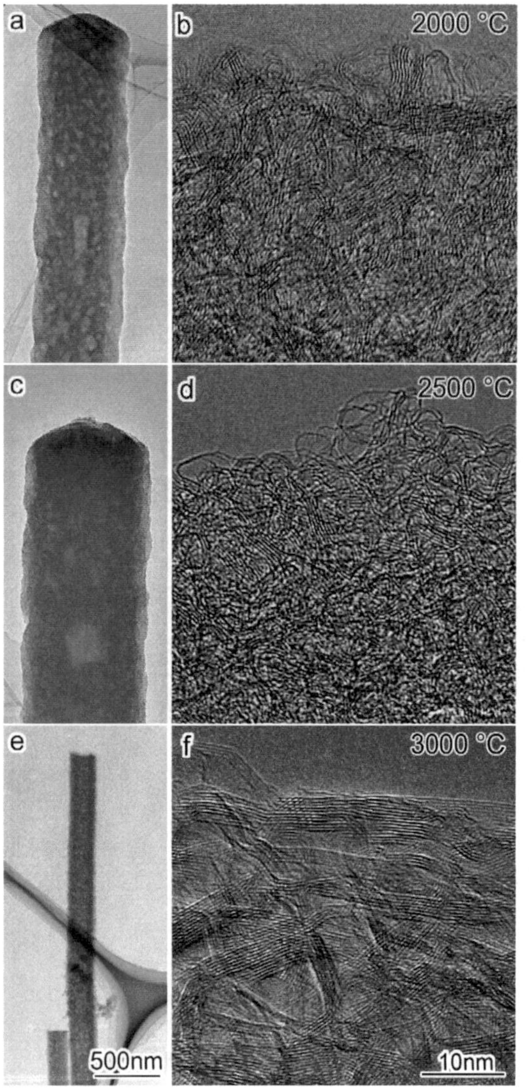

Figure 4. CTEM and high- resolution images of the C_{60} nanowhiskers heat-treated at 2000 °C (a, b), 2500 °C (c, d) and 3000 °C (e, f), respectively.

spherical shape of C_{60} was destroyed by heat treatment. The spectrum of the sample heated at 3000 °C is similar to that of glassy carbon [12-14].

Figure 4 shows low-magnification images and high-resolution images of the C_{60} nanowhiskers which were heated at 2000 °C, 2500 °C and 3000 °C, respectively. By the high-temperature heat treatment, the C_{60} nanowhiskers turned to glassy carbon nanowhiskers composed of randomly oriented ribbons with graphitic layers. The crystallinity of glassy carbon increases with

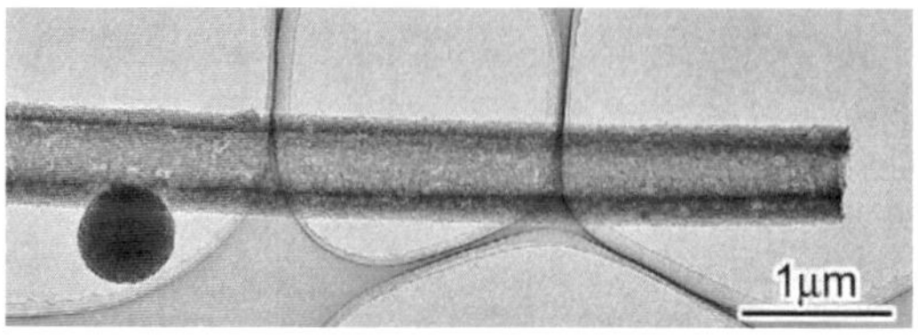

Figure 5. CTEM image of a C_{60} nanotube heated at 3000 °C.

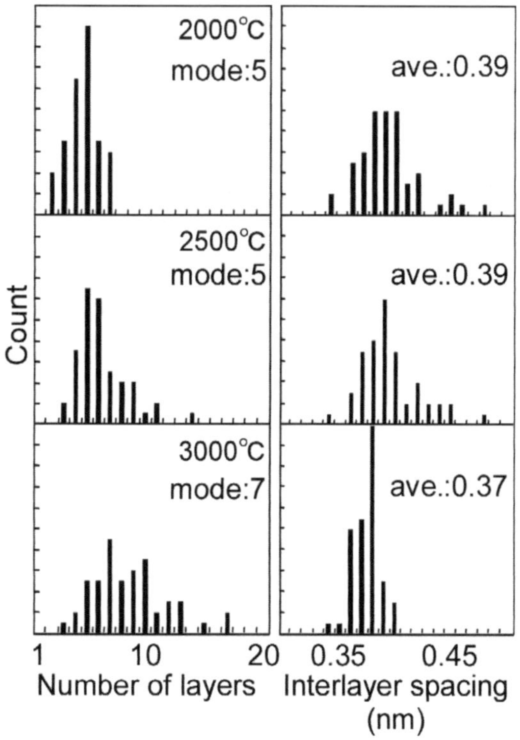

Figure 6. Histograms of the number of layers and the interlayer spacing (nm) of graphitic ribbons of the heat-treated C_{60} nanofibers.

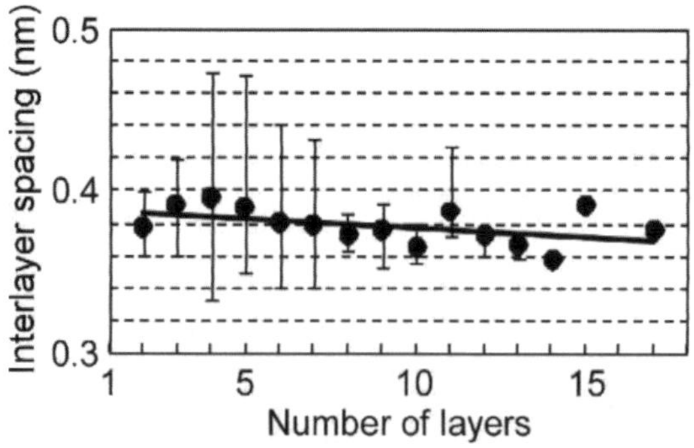

Figure 7. Relationship between the number of layers and the average interlayer spacing

increasing the temperature of heat-treatment. Therefore, the structure of nanowhiskers heated at high temperature became porous. However, the carbon nanocapsules which have fine circular shape couldn't be observed [6]. We think that a rapid increase of temperature is a factor to form the carbon nanocapsules.

Figure 5 shows a bright field TEM image of a C_{60} nanotube heated at 3000 °C. The heated nanotube retained their original tubular morphology. The inner structure of the heated nanotube is similar to the above heated nanowhisker. Owing to the tubular morphology, the C_{60} nanotubes must have a specific surface area higher than the C_{60} nanowhiskers. The tubular structure suggests that the C_{60} nanotubes heat-treated at high temperature can be utilized as a new electrode material.

Figure 6 shows the distribution of layer number and interlayer spacing of the graphitic ribbons of the C_{60} nanofibers heated at each temperature. The average interlayer spacings are close to the interlayer spacing of {002} planes of graphite. The maximum number of graphitic ribbons is counted to be 17 as shown Fig. 6. With decreasing the temperature of heat-treatment, the number of layers decreases and the average interlayer spacing increase as shown in Fig. 6.

Figure 7 shows the relationship between the number of layers (x) and the average interlayer spacing (y) as shown in Fig. 6. A linear relationship expressed by $y = -0.001\ x + 0.39$ is obtained. It is deduced that the interlayer spacing of heat-treated nanotubes with 49 layers converges to 0.3354 nm of {002} spacing of highly oriented pyrolytic graphite [15].

4. Conclusion

The above research can be summarized as follows.

(1) The C_{60} nanowhiskers and nanotubes became amorphous by the heat-treatment at 900 °C.

(2) The C_{60} nanowhiskers heated at temperatures higher than 2000 °C exhibited the structure of glassy carbon composed of randomly oriented graphitic ribbons with interlayer spacings close to {002} planes of graphite. The heat-treated C_{60} nanowhiskers became porous owing to the crystallinity development of graphitic ribbons.

(3) A linear relationship holds between the number of layers and the average interlayer spacing of graphitic ribbons of the heat-treated C_{60} nanowhiskers and nanotubes.

(4) The tubular morphology of C_{60} nanotubes can be retained by the heat treatment up to 3000 °C.

Acknowledgments
The authors are grateful to Mr. Hideo Tsunakawa (The University of Tokyo) for the use of HRTEM.

References

[1] Miyazawa K, Obayashi A and Kuwabara M 2001 *J. Am. Ceram. Soc.* **84** 3037
[2] Minato J and Miyazawa K 2006 *Diam. Relat. Mater.* **15** 1151
[3] Miyazawa K, Kuwasaki Y, Hamamoto K, Nagata S, Obayashi A and Kuwabara M 2003 *Surf. Interface Anal.* **35[1]** 117-120.
[4] Larsson M P, Kjelstrup-Hansen J and Lucyszyn S 2007 *ECS Transactions* **2** 27
[5] Miyazawa K, Minato J, Zhou H, Taguchi T, Honma I and Suga T 2006 *J.Eur.Ceram.Soc.* **26** 429-434.
[6] Asaka K, Kato R, Maezono Y, Yoshizaki R, Miyazawa K and Kizuka T 2006 *Appl. Phys. Lett.* **88** 051914-1-051914-3.
[7] Miyazawa K, Kuwasaki Y, Obayashi A and Kuwabara M 2002 *J.Mater.Res.* **17** 83-88

[8] Ringor C L and Miyazawa K 2008 *Diam. Relat. Mater.* **17** 529–534
[9] Miyazawa K and Ringor C 2008 Mater.Lett. **62** 410-413
[10] Martin M C, Koller D, Rosenbergand A, Kendzior C and Mihaly L 1995 Phys. Rev. B **51** 3210-3213
[11] Eklund P C, Zhou P, Wang K A, Dresselhaus G and Dresselhaus M S 1992 J. *Phys. Chem. Solids.* **53** 1391
[12] Diane S, Knight and William B White 1998 *J. Mater. Res.* **4** 385
[13] Robertson J 1986 *Advances in phys.* **35** 317
[14] Nakamizo M, Honda H and Inagaki M 1978 Carbon **16** 281
[15] Murakami M, Nishiki N, Nakamura K, Ehara J, Okada H, Kouzaki T, Watanabe K, Hoshi T and Yoshimura S 1992 *Carbon* **30** 255-262

Journal of Physics: Conference Series **159** (2009) 012025 doi:10.1088/1742-6596/159/1/012025

Auger Investigation of Fullerene Nanowhiskers Heat-treated at 400-800 Degrees Centigrade

M Harada[1] and D Fujita[2]

[1]Fukuoka University of Education, Munakata, Japan
[2]National Institute for Materials Science, Tsukuba, Japan

E-mail: haradab@fukuoka-edu.ac.jp

Abstract. Fullerene nanowhiskers were prepared from a toluene solution of C_{60} and isopropyl alcohol by a liquid-liquid precipitation method. The samples on the silicon (100) wafer were heated at the elevated temperatures ranging from 400 °C through 800 °C by an electron beam bombardment in an ultra-high vacuum, and their thermal stability was investigated with a scanning Auger-electron microprobe. At 400 °C fullerene nanowhiskers already got thinner, while bulky fullerene crystals still remained even at 800 °C. And many craters were observed around the fullerene crystals on the Si(100) substrate.

1. Introduction

Since the discovery of third carbon allotrope fullerene C_{60} [1], many fullerene families including carbon nanotubes [2] and endohedral metallofullerenes M@C_{60} [3] have been extensively investigated for their unusual properties. And various forms of C_{60} crystals were found to be generated from their liquid solutions. K. Miyazawa reported a fine fibrous structure of C_{60}, formed at room temperature [4]. Needle-like crystals of C_{60} precipitated from an interface between a toluene solution of C_{60} and isopropyl alcohol, and they typically grew more than 100 μm long with a sub-micron diameter, called nanowhiskers [5]. The C_{60} nanowhiskers are considered to exhibit a higher strength and a higher electrical conductivity through polymerization [6]. However the detailed properties of C_{60} nanowhiskers are still unknown and intensive studies are now under way. In this study the thermal stability of the C_{60} nanowhiskers at the elevated temperatures, ranging 400-800 degrees Centigrade, was investigated by scanning Auger-electron microprobe (SAM).

2. Experimental

2.1. Sample preparation of fullerene nanowhiskers

The C_{60} nanowhiskers were fabricated by a liquid-liquid interfacial precipitation method, as described [5]. A toluene (99.5 %, Wako Pure Chemical Industries, Ltd., Japan) solution of fullerene C_{60} (>99.9 %, Tokyo Kasei Kogyo Co. Ltd., Japan) powder with a concentration of 0.1 mass% was prepared. Solubility of C_{60} in toluene was reported 2.8 mg/mL (ca. 0.3 mass%) [7], and its concentration was less than saturation. This toluene solution of 2 mL was poured into a 5 mL glass bottle, and 2 mL isopropyl alcohol (99.5 %, Wako Pure Chemical Industries, Ltd., Japan) was added

[1] To whom any correspondence should be addressed.

Journal of Physics: Conference Series **159** (2009) 012025 doi:10.1088/1742-6596/159/1/012025

gently. Since the densities of toluene and isopropyl alcohol are 0.86 and 0.79 g/mL respectively, the isopropyl alcohol layer was formed over the toluene layer. Around the interface many seed crystals were generated. The sample bottle was kept at room temperature (approximately 23 °C) till they grew to be needlelike whiskers.

2.2. Thermal stability of fullerene nanowhiskers
Morphological and elemental analyses of the samples were performed with an equipment of scanning Auger microprobe (Ulvac PHI, SAM650). The C_{60} nanowhisker solution was dropped on the Si(100) wafer (p-type, 280 μm thick, Furuuchi Chemical Corporation), and used for SAM observations after its solvent promptly evaporated. To investigate the thermal stability of C_{60} nanowhiskers, the samples were heated at 400-800 °C stepwise by using a sample heater with electron beam bombardment. The heating temperature was monitored with an infrared radiation thermometer (CHINO Corporation). After the sample was heated at fixed temperatures (400, 500, 600, 700 and 800 °C for 10 minutes each) in the irradiation chamber, it was transferred into the measurement chamber. All measurements were done at room temperature under an ultra high vacuum of 10^{-7} Pa or better. During the heat treatment the pressure temporarily went up to 10^{-6} Pa.

3. Results and discussions
The scanning electron microscopic (SEM) observation was done for the samples prepared. Figure 1(a) shows a SEM image (5 kV, × 160) of the sample as prepared (that is, no heat treatment). There seem many C_{60} nanowhiskers entangled and particulate C_{60} crystals. The SEM images of the samples after heated at 400 - 800 °C for 10 minutes are shown in Figures 1(b)-(f). This result shows that most of C_{60} nanowhiskers already got thinner at 400 °C, and that only traces of C_{60} nanowhiskers can be seen after a heat treatment at 800 °C. On the other hand, the bulky C_{60} crystals gradually get smaller with an increase of the heating temperature, but many C_{60} crystals still remains at 800 °C. Since the remaining carbon materials are not determined as C_{60} crystals, strictly they should be written as the remnants of C_{60} crystals. This result means that the thermal stability of C_{60} nanowhiskers is considered lower than that of bulky C_{60} crystals.

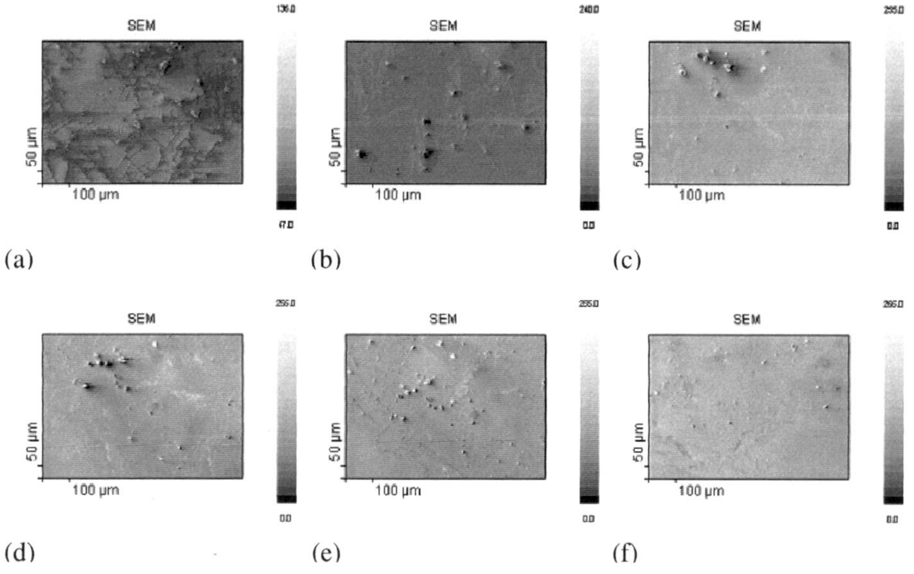

Figure 1. The SEM images of the C_{60} nanowhiskers (5 kV, × 160): (a) no heat treatment, and after the heat treatment at (b) 400 °C, (c) 500 °C, (d) 600 °C, (e) 700 °C, and (f) 800 °C for 10 minutes each.

Then the C_{60} nanowhisker samples were observed by scanning Auger microprobe for elemental analyses. The C_{60} nanowhiskers heated at 600 °C is shown in Figure 2(a) (5 kV, × 1000). There seem one large C_{60} crystal and two lines, the supposed traces of C_{60} nanowhiskers. Two points, one on a trace line of C_{60} nanowhisker and the other on the Si substrate, were measured by Auger. As shown in Figure 2(b), the Auger spectrum of the Si substrate presents silicon, oxygen, and carbon peaks, indicating the silicon oxide layer and the residual organic solvents. The spectrum of the trace line includes more carbon and less oxygen than the Si substrate, and so it means that the trace line must originate from the C_{60} nanowhisker. Such a framework of the C_{60} nanowhisker seems to be retained after a heat treatment at 800 °C, which may be a fullerene shell tube [8].

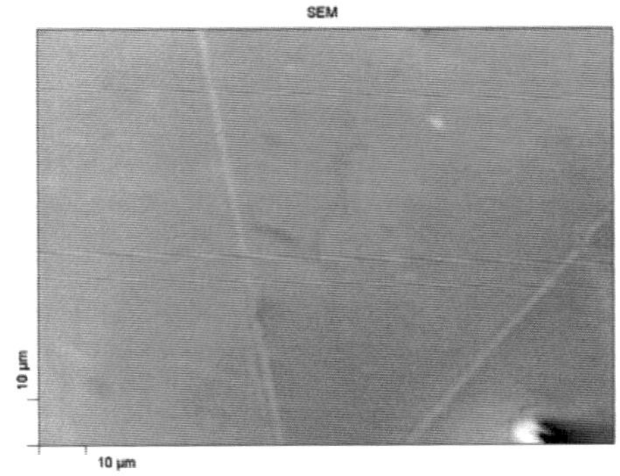

(a)

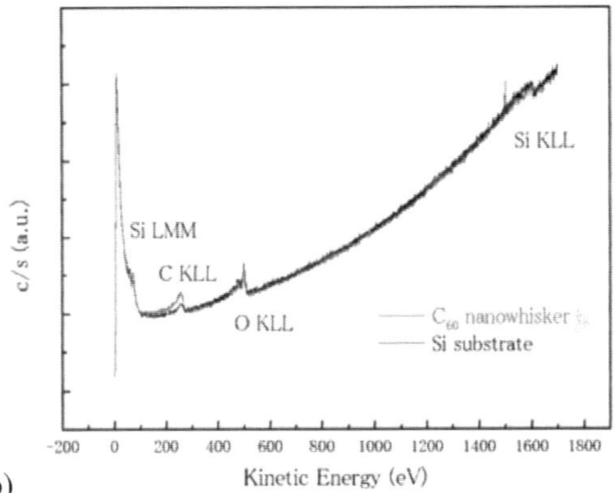

(b)

Figure 2 The C_{60} nanowhisker sample heated at 600 °C: (a) SEM image and (b) Auger spectra.

Figure 3 shows the Auger mapping of the large C_{60} crystals remained after a heat treatment at 800 °C, where (a), (b), and (c) represent the Auger maps of carbon (C KLL), oxygen (O KLL), and silicon (Si KLL), respectively (5 kV, × 2000). The large crystals mainly consist of carbon, while the substrate is composed of silicon and oxygen. Here it is noted again that the large C_{60} crystals are much more stable to the temperature than the C_{60} nanowhiskers. And there are also observed many craters on the Si substrate like a white spot in the Figure 3(c). Since the crater consists of silicon, the silicon oxide layer is considered to disappear anyhow. In fact the enlarged SEM image of this crater in Figure 4(a) confirms this. Their Auger spectra in Figure 4(b) show that the crater is the silicon (100) substrate itself and that the island observed in the middle of the crater is the remnant of the C_{60} crystal. Here it is noted again that the remnant carbon materials are not necessarily C_{60} crystals. The thermal decomposition of SiO_2 layer on Si were often reported, where the reaction of $SiO_2 + Si \rightarrow 2SiO \uparrow$ occurs [9,10]. But in our investigation a thickness of SiO_2 layer is estimated as thin as 2 nm by ellipsometry and the craters are usually observed with C_{60} crystals in the center. And so it suggests that the bulky C_{60} crystals play some role in a crater formation, but further study is to be done for clarifying the mechanism.

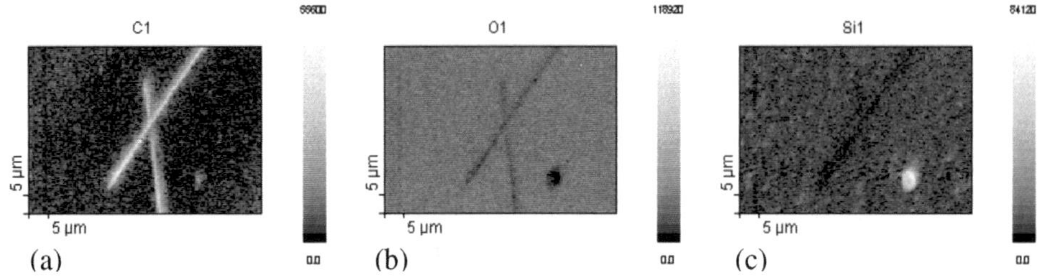

Figure 3 The Auger mapping of the C_{60} crystals heated at 800 °C: (a) carbon, (b) oxygen, and (c) silicon.

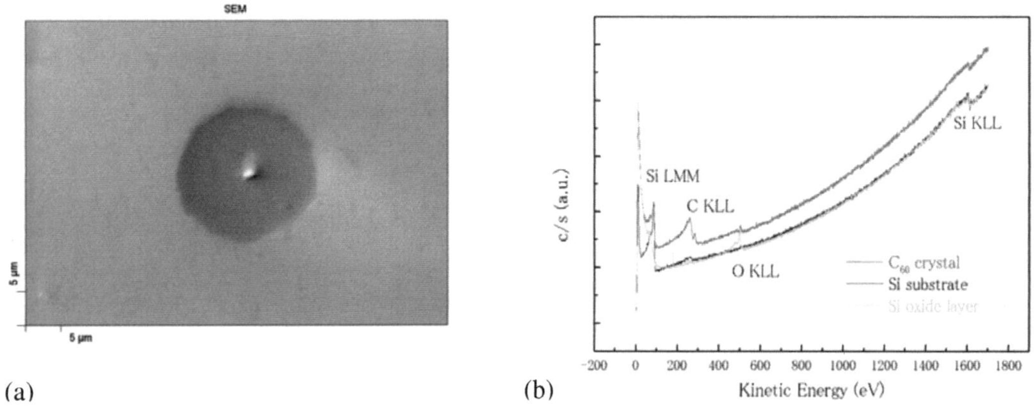

Figure 4 The crater on the Si(100) substrate: (a) an enlarged SEM image and (b) Auger spectra.

4. Conclusions

The C_{60} nanowhiskers and large crystals were formed with a liquid-liquid precipitation method. Their thermal stability is rather different. Almost all nanowhiskers already got thinner at 400 °C, while bulky C_{60} crystals still remained even at 800 °C. So the thermal stability of C_{60} nanowhiskers is lower than that of C_{60} crystals. And many craters were generated around the C_{60} crystals on the Si(100) substrate and it may be inferred that the C_{60} crystals should be a help towards the crater formation of the silicon oxide layer.

References

[1] Kroto H W, Heath J R, O'Brien S C, Curl R F, Smalley R E 1985 *Nature* **318** 162
Krätschmer W, Lamb L D, Fostiropoulos K, Huffman D R 1990 *Nature* **347** 354
[2] Iijima S 1991 *Nature* **354** 56
[3] Chai Y, Guo T, Jin C, Haufler R E, Chibante L P F, Fure J, Wang L, Alford J M, Smalley R E 1991 *J. Phys. Chem.* **95** 7564
[4] Miyazawa K, Obayashi A, Kuwabara M 2001 *J. Am. Ceram. Soc.* **84** 3037
[5] Miyazawa K, Kuwasaki Y, Obayashi A, Kuwabara M 2002 *J. Mater. Res.* **17** 83
[6] Miyazawa K, Kuwasaki Y, Hamamoto K, Nagata S, Obayashi A, Kuwabara M 2003 *Surf. Interface Anal.* **35** 117
[7] Ruoff R S, Tse D S, Malhotra R, Lorents D C 1993 *J. Phys. Chem.* **97** 3379
[8] Miyazawa K, Minato J, Zhou H, Taguchi T, Honma I, Suga T 2006 *J. Euro. Ceram. Soc.* **26** 429
[9] Tromp R, Rubloff G W, Balk P, LeGoues F K 1985 *Phys. Rev. Lett.* **18** 2332 (1985)
[10] Liehr M, Lewis J E, Rubloff G W 1987 *J. Vac. Sci. Tech. A* **5** 1559

AUTHOR INDEX